The Crafty Kid's Guide to DIY Electronics

About the Author

Helen Leigh is an author, education writer, and maker with a focus on creative use of new technologies. She has written playful technology education materials for *National Geographic*, Intel Education, and Adafruit and has developed a Design, Coding, and Electronics Course for the Royal Court of Oman.

Alongside her writing, Helen makes creative technology products with a focus on education, including her latest collaboration with Imogen Heap, MI.MU, and Pimoroni, a gesture-controlled musical instrument glove that you can sew, wire, code, and play. To see some of the things Helen has made and find out more about some of the projects she has worked on, visit her website, www.doitkits.com.

Helen lectures on electronics, physical computing, and music technology at Ravensbourne University and Tileyard Studios in London. She was previously director of the education platform Mission:Explore, with whom she published six acclaimed children's books.

Helen lives in Berlin but is often found in London. You can say hello, ask questions, or show off your DIY electronics on Twitter (@helenleigh), on YouTube (HelenLeigh), or on Instagram (@helenleigh_makes).

The Crafty Kid's Guide to DIY Electronics

20 Fun Projects for Makers, Crafters, and Everyone in Between

Helen Leigh

New York Chicago San Francisco Athens London
Madrid Mexico City Milan New Delhi
Singapore Sydney Toronto

Library of Congress Control Number: 2018954683

McGraw-Hill Education books are available at special quantity discounts to use as premiums and sales promotions or for use in corporate training programs. To contact a representative, please visit the Contact Us page at www.mhprofessional.com.

The Crafty Kid's Guide to DIY Electronics: 20 Fun Projects for Makers, Crafters, and Everyone in Between

1 2 3 4 5 6 7 8 9 LOV 23 22 21 20 19 18

ISBN 978-1-260-14283-9
MHID 1-260-14283-3

Sponsoring Editor
Lara Zoble

Editing Supervisor
Donna M. Martone

Production Supervisor
Lynn M. Messina

Acquisitions Coordinator
Elizabeth Houde

Project Manager
Patricia Wallenburg

Proofreader
Claire Splan

Indexer
Claire Splan

Art Director, Cover
Jeff Weeks

Composition
TypeWriting

For my sister, Nicola Picola,

despite asking me "How's the book going?" way too often.

Contents

Foreword

Our imaginations drive us all through our lives. We spend our days learning and our nights dreaming. We learn how to walk, how to run, how to dance, how to add and subtract, and how to read. We go to school to learn, and we learn from the wider world outside. We learn when we work and, perhaps more importantly, we learn when we play. The things we learn, the stories we tell, the art we make—these things all mesh together to help us understand our lives and let us express who we are to the world.

Inside each of us there is a creative spark, a little fire that burns a bit brighter every time we learn something new that inspires us. This book is all about creative technology, about using design, art, fashion, music, science, and play all together. It's about making things that can be beautiful and practical, as well as fun and silly. It's about taking a needle and thread and mapping out a patchwork constellation to shine above you at night. It's about telling your own stories, creating your own practical arts and crafts and learning how to stitch together your own magic.

Everything you're going to read about here in this book is something you can already do. Helen won't sit you down and dictate things to you; she wants to show you how to look at things in a new way for yourself. Her ideas are all about taking one skill, one way of expressing yourself—whether it's sewing or origami or drawing—and building on that. But more importantly, her ideas are about building up that little fire that was already there inside you and giving you the confidence and the help to look at technology and invent the future yourself.

In this book you'll find out how to make a pair of magic gloves you share with your best friend, a robot that wakes you up in the morning, and a paper skyline that looks like your own home town at night—or even a city that never existed, except in your own imagination. It's so exciting to think of you taking your first steps into the creative world of electronics.

I can't wait to see what you make.

Imogen Heap xx

Imogen Heap is a Grammy award–winning musician and innovator.

Introduction

Hello.

Welcome to *The Crafty Kid's Guide to DIY Electronics*. My name is Helen and I wrote this book. As well as being a writer, I like to invent and make things. Some of my inventions have been quite serious and scientific. Some of my inventions have been very silly and full of joy. I've made extremely high-tech devices that talk to the Internet and measure pollution, I've made lots of electronic and analog instruments and I've also made herds of gesture-controlled robot unicorns for children (and very excited adults) to race.

I think that technology is awesome, but I also think that playfulness, creativity, and craft are just as important. I believe that when we mix up two different skills or interests we make them both better—and we look at the things we thought we knew before in a different way. In this book I want to show you some of the super-cool things you can make when you combine craft skills with electronics know-how.

Even though I want to share some of the things I've learned with you, this book is not a textbook. You won't find any formulas or lengthy lectures about electrical engineering. Don't get me wrong, there are loads of great facts and explanations in this book—and the science of electricity is super-interesting—but that's not what this book is for. *The Crafty Kid's Guide to DIY Electronics* is about making cool things, exploring new electronics and craft concepts, and getting hands-on to develop the skills you'll need to invent your own projects in the future.

When I talk to young people about the work I do, I always get asked the same kind of questions. I get asked about my favorite invention (probably the robot unicorns), I get asked if I've ever blown things up or set them on fire (yes, all the time, and I once set my eyebrows on fire too), and I get asked what a child should learn if they want to grow up and be an inventor or a maker. The answer to this last question is surprising to most people. There are three important things that I've learned over the course of my life that help me do what I think is the best job in the world.

Thing Number One: Resilience

In this book we're going to explore lots of different ways of making and inventing cool things. You might find some of the new techniques and concepts difficult or confusing at first. You'll almost certainly get something wrong and have to undo your work or worse—start over again from scratch. I know what that feels like, and it's not nice. Resilience is the ability to bounce back from this feeling quickly. I never get things right the first time I make something—making mistakes and trying again is essential to being a good inventor and maker.

Thing Number Two: Creativity

I think of creativity as applied imagination. One of the most famous scientists of all time, Albert Einstein, said that "imagination is more important than knowledge," and I agree. Our daydreams, our thoughts, the stories we tell ourselves, and the places we go in our minds shape who we are in every way. Creativity is when we take little pieces of our imagination and make them real. To be an inventor you'll need to practice taking impossible, impractical, and beautiful ideas out of your mind and into the world.

Thing Number Three: Practical Skills

A skill is knowing how to do a thing or make a thing work. That thing could be threading a needle, it could be using a tool the proper way, it could be knowing how to use code to make a robot move, or it could be knowing what materials to use in different situations. As an inventor you'll need to collect a lot of different skills and you'll also have to learn where—and who—to ask for help.

While you read *The Crafty Kid's Guide to DIY Electronics* I want you to keep these three things in mind. If something goes wrong, practice your resilience by keeping a cool head, working out what the problem is, and trying again. I'd also like you to practice your creativity. In this book I've shared 20 things I've made, but I hope you don't make them in exactly the same way as me. In many of the projects I explain how I came up with my ideas and I hope this inspires you to come up with your own twists on my makes. I'll also be sharing loads of useful practical skills along the way, including how to sew different types of stitches, how to strip electrical wires, how to tie useful knots, and how to check that your electricity is going where it should.

Next, I'll lay out the things you'll learn, the tools you'll need and the materials you'll use in more detail, but first I want to talk to you about where to look and how to ask for help when you need it. I lecture on electronics at a university in London and I always joke that the most important thing that I teach my students is how to use a search engine properly. We live in a time where anyone with access to the Internet has an amazing opportunity to learn how to do pretty much anything. The problem is finding it. There are a number of great websites I go to for good quality tutorials and advice. My go-to websites for electronics guidance and inspiration are Adafruit, Sparkfun, Instructables, KobaKant, and YouTube.

Finally, I want you to feel like you can ask me for help. I am active on Twitter and help people troubleshoot their projects almost as often as I ask other people for help with my own projects. I even have a YouTube channel of my own, where I post videos of things I make and simple instructions on how to use tools. You'll find step-by-step videos of many of the makes in this book and you can post comments to ask for guidance if you get stuck.

Find me at www.youtube.com/c/HelenLeigh or www.twitter.com/HelenLeigh.

How This Book Works

This book has four parts that focus on different types of crafty projects: Paper Circuits (page 3), Soft Circuits (page 73), Wearables (page 137) and Robots (page 197). In each part you'll find a range of step-by-step projects, from entry-level makes that teach basic techniques to more complex challenges that encourage you to level up your electronics and craft skills. Each project introduces at least one new skill or concept, which you can find by looking for the owls.

The owls signal new knowledge. You can read these sections in the context of the make, but you can also read them on their own. At the beginning of each part you will find a handy list of skills and concepts covered in each project, with a page number so you can find them easily.

What You'll Need

When I started writing this book I thought very carefully about the kind of tools and materials I wanted to use. I tried to keep each project as low cost as possible, with lots of reusable materials across the different parts. I hope you'll have most of the tools and craft materials at home already, but if you don't they are easy to find and cheap to buy. If you don't already have one I would suggest investing in a good, sharp craft knife (sharper is safer, as you'll find out on page 60) and a small cutting mat.

There are some materials that you probably don't already have. These are the electronics components. You can find them easily online. My four favorite shops are Pimoroni, Adafruit, Kitronik, and Sparkfun. Again, I've tried to keep the cost to a minimum. All the components you buy will be reusable in other projects—either my projects or your own inventions.

To make all the projects in this book, you'll need the following tools.

- One 12-inch ruler, ideally metal with a non-slip back
- Flexible tape measure
- Scissors
- Needle-nosed pliers
- A sharp craft knife
- Wire cutters
- Sewing needles
- Sewing pins
- Embroidery hoops
- Cutting mat
- Clear nail varnish
- Hot glue gun (optional)
- You'll also need the following materials.

From a Craft Store

- Assorted colors of card
- Assorted colors of paper, including white and black
- Assorted colors of origami paper, including red
- Crepe or tissue paper in any color
- Scrap cardboard
- Fifteen 8-inch squares of soft felt, including pink, yellow, and navy
- Assorted colors of ordinary cotton thread, including pink and yellow
- Assorted colors of embroidery thread, including white
- Buttons, beads, and pom-poms
- Wadding
- Craft glue
- Sticky tape
- Pegs or small bulldog clips
- Two pairs of metal press studs
- Paper- or plastic-covered bendable craft wire
- Leather or nylon cord
- One sewable badge pin
- One headband
- Two pairs of woolen gloves
- Googly eyes

From an Electronics Store

- 5mm wide copper tape roll
- Ten Chibitronics LED stickers
- White and yellow 3mm or 5mm LEDs (I bought a pack of assorted LEDs in a nice little box from Amazon)
- Two sewable/stickable buzzers (I bought mine from Kitronik)
- At least ten sewable LEDs (I bought mine from Kitronik and Adafruit)
- Three sewable Lilypad or Adafruit LED sequins in different colors
- One sewable on/off switch
- One sewable/stickable vibration motor (I bought mine from Teknikio)
- One Teknikio star sewable LED
- Two Teknikio heart sewable LEDs
- One Teknikio RGB LED
- One Teknikio sewable light sensor
- One reel of conductive thread (I bought mine from Adafruit and Pimoroni)
- One square of conductive fabric (I used EeonTex from Sparkfun)
- Stranded electrical wire
- At least one 3V DC motor
- At least two sewable 3V battery packs with on/off switches (I bought mine from Pimoroni)
- 3V batteries

Troubleshooting

At the end of each project you'll find my guide to troubleshooting. As I said on the last page, you'll probably get things wrong quite often. When something doesn't work, practice your resilience by keeping a cool head and going through my troubleshooting guide step by step. I'm really good at finding out what's wrong in other people's projects, and it's not because I'm some amazing electronics genius—I simply have a method that I work through.

First, I check the **power**. Is the battery in? Is it the right way round? Is it switched on? Is it flat? You'd be astonished at how many complex projects don't work for one of these simple reasons.

Next, I check the **components**. Are they the right way round? Are they broken? Are they connected in properly? Checking each component before you add it into your circuit is a great habit to get into.

Finally, I check the **connections**. Are my paths going where they should? Is there a short circuit? Is there a tear in my copper tape or did I accidentally snip off a corner? Trace the path with your finger to make sure your electricity is going where you want it to. Remember, electricity is lazy. If you give it the chance to avoid an obstacle, such as a component, it will. Even a tiny connecting bit of tape or thread will stop your circuit from working.

Work through each of these in order and you'll almost always figure out what's wrong on your own. It's not a nice feeling when your project doesn't work, but it's an *amazing* feeling when you figure out why and fix it yourself.

I hope you enjoy this book as much as I enjoyed writing it. I am so excited to share my ideas with you. Let's get making!

Yours,

Helen Leigh xx

www.youtube.com/c/HelenLeigh
www.twitter.com/HelenLeigh

Acknowledgments

In February 2018 I sent out a tweet asking if any of my friends had children that might be interested in helping me out with this book. Hundreds of people from all over the world replied, including families in Australia, teachers in France, librarians in the USA, and Code Clubs in the North of England.

As a result of that tweet, two hundred girls between the ages of seven and fourteen joined my youth advisory board and I want to express my gratitude to every single one of them. You have helped me with every major decision I've made, from choosing the title of this book to voting on what projects to include. This book would have been completely different without your help, advice, and encouragement. Thank you.

I am so lucky to be a little part of the maker communities in London, Berlin, Shenzhen, and on Twitter. Makers are a group of people who will always find time to help, teach, laugh or commiserate. Big love to Rehana Al-Soltane, Johannes Lohbihler, Rachel Freire, Ross Atkins, Hadeel Ayoub, Rachel Wong, Lit Liao, Lorraine Underwood, Paul Beech, Tanya Fish, Mike Barela, Jill Hodges, David Whale, Mitch Altman, Tom Fox, Ben James Simpson, and Carrie Leung. A special thank you to Colleen Jordan and Simon Monk for being so kind and for writing the books that I love.

I'd also like to thank all the makerspaces in London and Berlin where the projects in the book were prototyped and made. The old warehouse on the Regents Canal in Hackney that housed Machines Room is now an office, but for a few beautiful years it was the best place in the world. Nat, Mark, and Gareth, thank you for all your hard work. It was special and worthwhile.

All the love in my heart goes out to my wonderful sister, Nicola, and your rad husband, Dave. You're my rock and I am so excited to meet baby Ian next year. Endless thanks to Mum and Doug and Dad and Liz for your support and not being too grumpy when I forget to call. Thank you for teaching me and Nicola to not be scared of power tools, technology, bruised knees, or hard work.

My friends are amazing. Phoenix, you inspire me to become a better maker and a more radical human being. Thank you for your friendship and for showing me how to use an oscilloscope. Brett, you missed out on a lot of good times to help me get through writing this book. Thank you for constantly (sometimes annoyingly) reminding me to keep it focused on the fun. I can't wait to see what you do next. Huge thank you to my writing buddy Martin for your encouragement and suggestions. Big love to Cathy, Tammy, Tasha, and Katie. Emily, Hazel, and Emma, thank you for your love and gifs. Here's to the next 20 years.

Thank you to Michael McCabe and the rest of the McGraw Hill team for enthusiastically saying yes to me writing a book filled with my slightly odd ideas. Elizabeth, you rock. Patty, thank you for your kindness and hard work getting this book over the finish line.

PART ONE

Paper Circuits

Introduction to Paper Circuits

Welcome to the Paper Circuits part. We're going to get crafty with electricity, making interactive paper and cardboard projects that light up, buzz, vibrate, and dance.

This part will introduce you to the world of electricity and circuits. By the time you finish this part, you will have the skills and knowledge you need to design and make your own paper circuit projects.

Over the course of this part, I'm going to guide you through the basics of paper circuitry. I'll give you lots of tips and tricks for using interactive components in your craft projects. On the next page, I'll explain some of the basic facts about electricity that you'll need to put into practice while making your paper circuits. Later in this part, you'll learn how to avoid basic mistakes and how to design, make, and power your projects. Each of these skills forms part of one of the five projects in this part, but if you want to skip ahead or look back, here's where to find this information:

Paper Circuits: Essential Skills

- How to work with copper tape (pages 10–12)
- How to work with cardboard (page 31)
- How to plan more complex paper circuits (page 62)

We will also start to explore the world of DIY electronics. By the end of this part, you will understand what a circuit is, how to guide electricity on a path, how to power your projects, and how to use some common electronics components such as light-emitting diodes (LEDs), buttons, and buzzers. Here's where to find this information:

- What is an LED sticker? (page 10)
- Adding more than one LED (page 12)
- Introducing series circuits (page 13)
- Introducing parallel circuits (page 13)

- What are buttons and switches? (page 32)
- How to move electricity along a path (page 34)
- How to make a parallel circuit (page 43)
- How to use an LED in a paper circuit (page 52)

This part will also challenge you to try lots of new craft skills. You'll learn lots of techniques and tricks to take your paper craft projects to the next level. Here's where you can find all this information:

- How to make your first origami fold (page 21)
- How to work with cardboard (page 31)
- How to score card (page 42)
- How to make a reverse fold (page 50)
- How to make a pop-up card (page 59)
- How to use a craft knife (pages 60–61)

Paper Circuits: Essential Skills

What Is Electricity?

Humans use electricity to power all sorts of things, from the televisions in our homes to complex science experiments on the space station. All these uses of electricity are based on the same simple ideas that we'll be exploring in this book.

Electricity is made up of a flow of tiny particles called *electrons*. In order for electricity to do anything, these electrons must have a clear path to follow. We can make these electrons do some pretty cool things by carefully controlling their flow and putting obstacles in their path. This path is what we call a *circuit*.

What Is a Circuit?

A circuit is the path electricity takes. For a circuit to work, it needs a power source, such as a battery. Take a look at a standard AA battery. You will see a + (positive) symbol at one end and a – (negative) symbol at the other.

We can make a path for the electricity inside the battery by connecting the positive and negative sides to a light, with wires connected to both the positive (+) and negative (–) ends. A circuit can also contain other things, including buzzers, sensors, and motors, which all do cool things when you pass electricity through them. We call these things *components*.

A circuit will not work if the path is broken. Electricity will only travel around a complete circuit with no gaps. Look at the following diagram to see a complete circuit and an incomplete circuit.

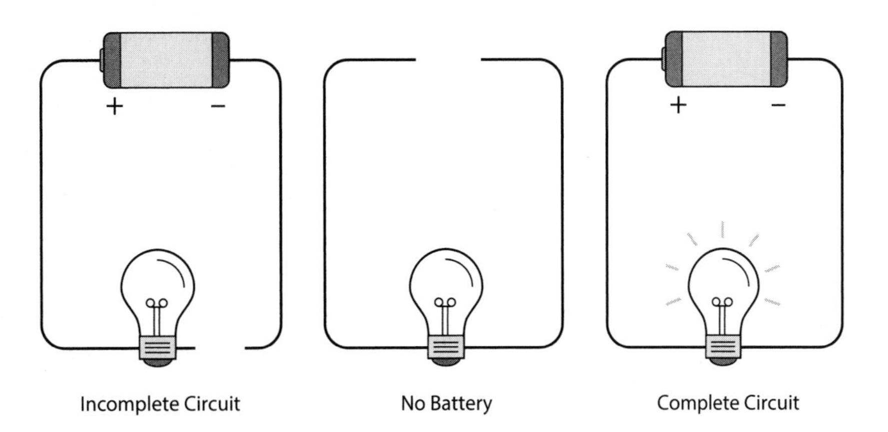

Conductive and Nonconductive Materials

When most people think of making a circuit, they think about getting the electricity to move through wires. However, you don't have to use wires to make a circuit—you can use anything that is conductive. When we say something is *conductive*, we mean that electricity can pass through it. When we say something is *nonconductive*, we mean that it will block the path of electricity.

Conductive materials include copper tape, special fabrics, graphite (what we call *lead* in many pencils), and even some types of sewing thread. Nonconductive materials include paper, card, parcel tape, and plastic.

Buying Copper Tape

You can buy copper foil tape in a few different thicknesses. It comes in two types: with conductive glue or with nonconductive glue. I usually use 5-mm copper tape with conductive glue, but the techniques we are going to learn will work with any type of copper tape.

You can order copper tape with conductive glue online. Copper tape with nonconductive glue is more common. You can find it very cheaply at hardware and gardening stores, labeled as "slug tape."

Quick Start Tips

You'll find lots of good tips and tricks for crafting your paper circuits as you read through this part. Here are the most important things to remember.

- *Connect the positive bit of the battery to the positive bit of your component and the negative bit of the battery to the negative bit of your component.*
- *Electricity likes to take the easiest route possible. If you put a component on top of an unbroken bit of copper tape, the electricity will go through the copper tape instead of the component.*
- *Make sure your positive and negative paths do not touch. If they touch, your circuit will not work.*

Tools and Materials List

To make all the projects in this chapter, you'll need the following tools:

- Sharp scissors
- Craft knife
- Cutting mat
- Ruler, ideally metal with a non-slip back

You'll also need the following materials:

From a Craft Store

- Assorted colors of card
- Assorted colors of paper, including white and black
- Cardboard
- Assorted colors of origami paper, including red
- Craft glue
- Sticky tape
- Pegs or small bulldog clips
- Picture frame

From an Electronics Store

- 5-mm-thick copper tape roll
- Ten Chibitronics LED stickers
- One sewable/stickable buzzer
- One sewable/stickable vibration motor
- Six 3V batteries

PROJECT 1

Light-Up Greeting Card

Get crafty with electricity by designing and making your own light-up greeting card

Tools

- Sharp scissors
- Ruler, ideally metal

Materials

- Card
- Paper
- Glue
- Copper tape
- Two Chibitronics LED stickers
- 3V battery
- Wooden clothespin or small bulldog clip

Lots of people love receiving cards, and getting a homemade card feels even better. Taking the time to design and make a card shows that you really care about someone.

In this project, we'll be taking our first steps in DIY electronics. You'll make your first paper circuit, and you'll design and make your first unique project. By using your new electronics skills to add electricity and light to your greeting card, you can really blow your friends' and family's minds.

Let's make!

Choosing Your Materials

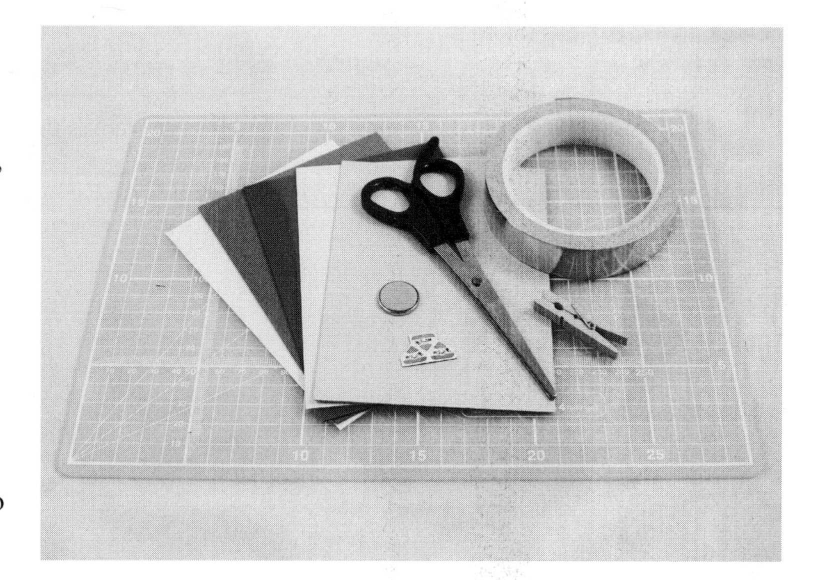

Start thinking about what you want your finished greeting card to look like. Before you dive into design, think about who you want to give it to. What occasion will it mark? You could choose to make one for Mother's Day, Christmas, Hanukkah, or your best friend's birthday.

Once you've decided who and what your card is for, you can start to come up with ideas for the design. I've made a birthday card for my friend Adam. He loves cupcakes, so I want to make him a cupcake-inspired card with a candle that lights up. I've chosen a bright blue card that will contrast nicely with the copper tape.

Once you've thought about the way you want your card to look, prepare your materials. Make sure that you've read the "Paper Circuits: Essential Skills" on pages 4–6 before going any further.

Starting Your First Circuit

Before you start thinking about how you're going to design your card's circuit, we are going to build a simple practice circuit on a piece of paper. We'll be using three components: copper tape, a Chibitronics LED sticker, and a 3V battery.

Copper tape is conductive, which means that it can be used to make a path for electricity. We'll be using it to carry the electricity around the circuit from the battery to the Chibitronics LED sticker and back.

Our 3V battery is the source of power we need to make our circuit work. Take a look at your 3V battery. Flip it over, and you should be able to see that one side of the battery has writing on it, including a + symbol. This is the positive side of the battery. The other side is usually bumpy. This is the negative side of the battery. We need to connect our circuit to both the positive and the negative sides of the battery.

On a piece of paper, trace the template circuit on page 247. This is the layout of your first circuit. The dotted corner should be at the edge of your paper because we'll be folding it along the dots to make contact with the positive and negative sides of the 3V battery. The triangle-shaped space is for your Chibitronics LED sticker.

What Is an LED Sticker?

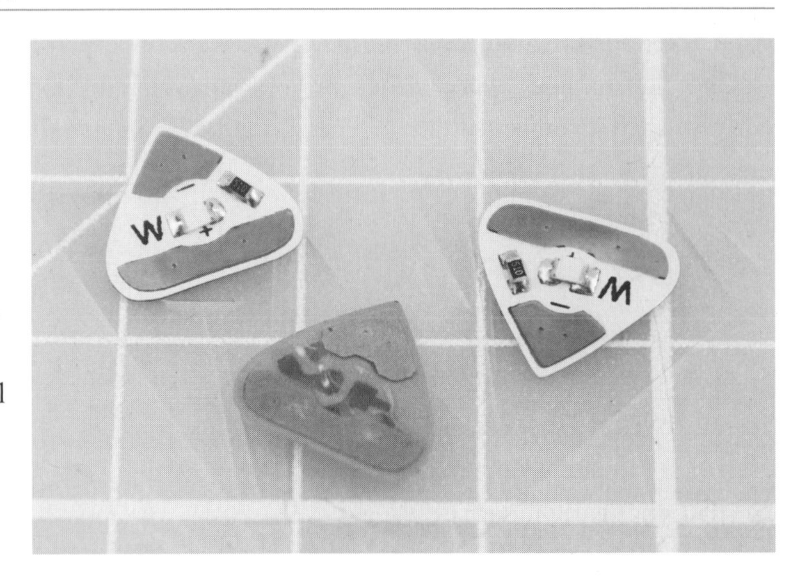

Take a closer look at your LED sticker. On the top of each sticker you can see a tiny yellow bump. This bump is an LED, short for light-emitting diode. An LED is like a tiny light bulb. If you put this LED sticker into a circuit correctly, it will light up!

You will also see two shiny strips on your sticker. The long thin strip is positive, and the short, pointed strip is negative. If you look carefully, you'll be able to see the + and – symbols. You will need to match these symbols up with your battery's positive and negative sides to make your circuit light up.

These awesome stickers were designed by Jie Qi of Chibitronics, an engineer who loves to craft with paper. Chibitronics has made some great videos to help you use copper tape and LED stickers. Search on YouTube for "getting started with copper tape for paper electronics" to find a great starter guide with lots of tips and tricks.

How to Work with Copper Tape

Peel the backing off your copper tape, and stick the tape down where the path should go, making sure to leave gaps in the tape where your components will sit. Remove the backing bit by bit to keep the length of tape from sticking to itself.

Try to use one continuous piece of copper tape between components. This will give you the most reliable connection. If you run out of tape before finishing your circuit, you can connect two bits of tape together.

If you need to connect two bits of copper tape, don't just stick one bit on top of the other. Even if your glue is conductive, this can make an unreliable connection, meaning that your LED might flicker or not work at all. Instead, fold under about half an inch at the end of the new bit of tape so that it sticks to itself. Put this bit of tape over the old bit of tape, and then stick it firmly in place with regular sticky tape.

To turn a corner, turn your tape over in the opposite way to the path you want to take, just as in the picture (on the previous page). The sticky side of your tape should be face up. Carefully flatten the turn. Next, turn the tape back over in the direction that you want it to go. Carefully flatten the corner again, and off you go! Corners can be tricky at first. Practice until you get the hang of it.

Laying Down Your Paths

Cut off a piece of copper tape about 8 inches long. Using your traced pattern as a guide, peel back one of the edges of the backing of your copper tape, and start sticking it down along your positive path. Remove the backing bit by bit.

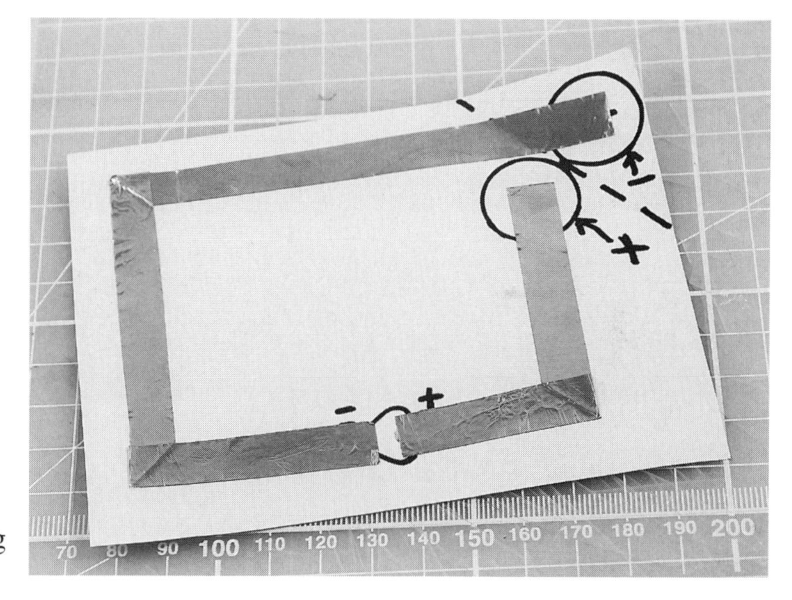

When you get to the first corner, carefully pull the tape over on itself so that it is sticky side up and pointing in the opposite direction to the path. Flatten the turn, and then pull the tape in the other direction so that the sticky side is down and your tape is pointing in the direction of the path. Flatten this turn too, and then keep sticking your tape to your path until you get to the space for the LED sticker.

Tear your copper tape so that it ends in the space for the LED sticker. Then repeat the process for the negative side of your circuit. Make sure that the positive and negative bits of copper tape are inside the space for the sticker but not touching each other; otherwise, your circuit won't work.

Finishing Your First Circuit

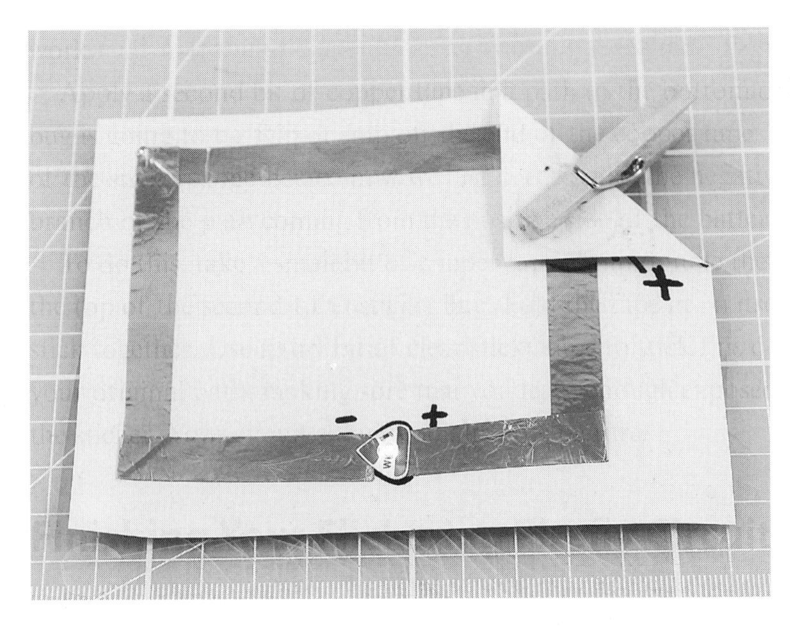

Crease the corner of the paper along the dotted line, and place the battery negative (–) side up over the positive (+) circle. Fold the corner flap over, and clip the battery in place with a wooden clothespin or a bulldog clip.

Make sure the positive and negative paths made of copper tape are not touching each other. Then stick the LED circuit sticker on top of the copper tape over the triangular space. Pay attention to the way in which you stick your sticker. The positive side of the LED sticker should connect to the positive side of the battery, and the negative side of the LED sticker should connect to the negative side of the battery.

When your sticker is in place, it should light up. Success? Give yourself a round of applause because you have just built your first paper circuit.

Not working yet? Skip to the end of this project for my troubleshooting guide.

Adding in More than One LED

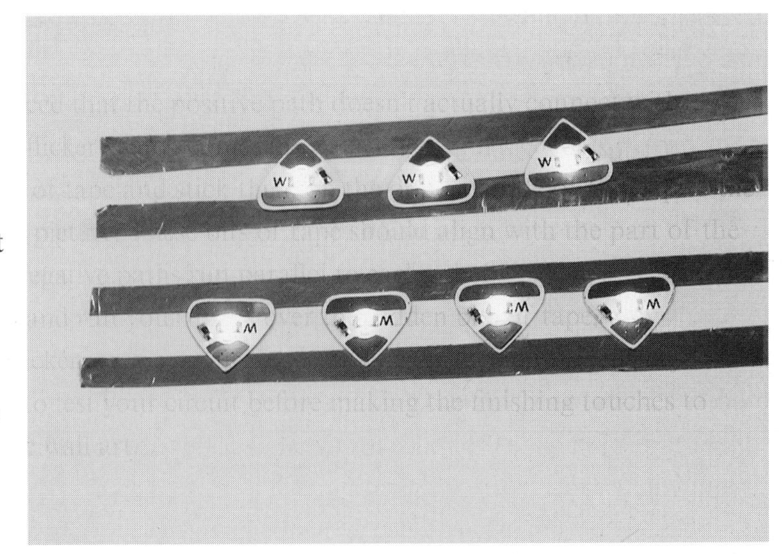

Making one LED sticker light up is awesome, but what if you want to make a design with more than one LED?

When you design a project with more than one LED, there are two ways to do it. The first way is to connect your LEDs in series, and the second is to connect them in parallel. We'll use both types of circuits later in this book, but for now, please read the following sections for an introduction.

Each LED sticker takes a certain amount of power from the battery, so if you want to add more than one sticker, you'll have to know more about how to balance power in your circuit.

It may be tempting to design a circuit with a whole pack of shiny LED stickers, but for your first project you should stick to designing a project with one sticker. By the end of the "Paper Circuits" part of this book, you'll be a pro at using several components!

Introducing Series Circuits

LEDs connected in series are connected end to end, meaning that the negative bit of the first LED connects to the positive bit of the second LED, the negative bit of the second LED connects to the positive bit of the third LED, and so on.

This whole daisy chain of LEDs should be connected up as usual to your power supply: negative to negative and positive to positive. In a series circuit, if one component is faulty, no part of the circuit will work.

The diagram shows a simple series circuit.

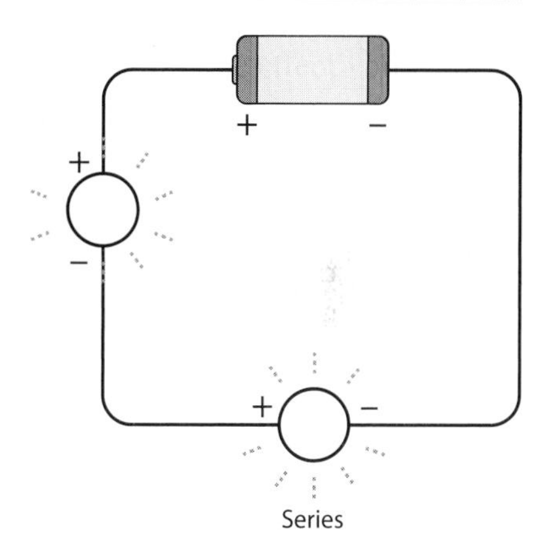

Series

Introducing Parallel Circuits

Another way we can make a project with more than one LED is to connect them in a parallel circuit. LEDs connected in a parallel circuit use one path to connect all the positive bits of the LEDs to the positive bit of the battery and another path to connect all the negative bits of the LEDs to the negative bit of the battery.

In a parallel circuit, if one component is faulty, the other parts of the circuit will still work. A circuit with two LEDs and one battery connected in parallel will make your LEDs shine more brightly than the same circuit connected in series, but it will also drain your batteries more quickly.

The diagram shows a simple parallel circuit.

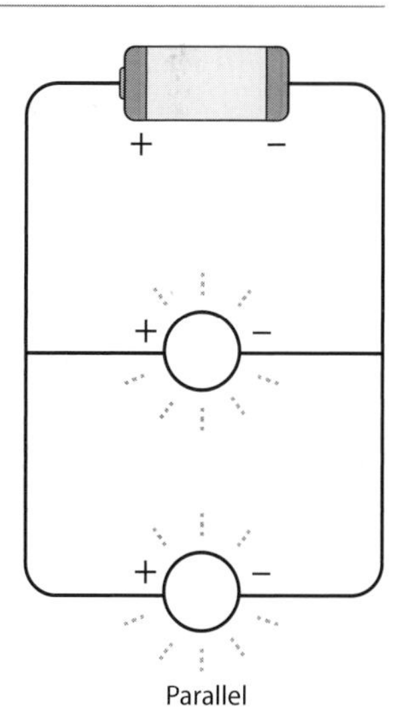

Parallel

Designing the Circuit for Your Greeting Card

Now that you know the fundamentals of paper circuitry, you can get start by creating your own greeting card design.

Sketch out some ideas for your design, starting with the circuit. Keep it simple to start with, using just one sticker and keeping corners to a minimum.

Clearly mark up your negative and positive paths and how they will connect to your battery. I'd suggest putting the battery in the bottom corner of the front of the card, but you could take the paths round to the back of the card if you'd prefer to hide your battery.

Think about how the rest of your design will work with your circuit. My design uses one LED as the flame on a birthday candle, but you could make a star on top of a tree, a light in the middle of a flower, or a glowing heart. Use your imagination. What can you light up?

You can also choose to cover your LED sticker with some thin paper so that the LED shines through, creating a pretty illumination effect. Don't do this with card or your LED won't show through.

Making the Circuit for Your Greeting Card

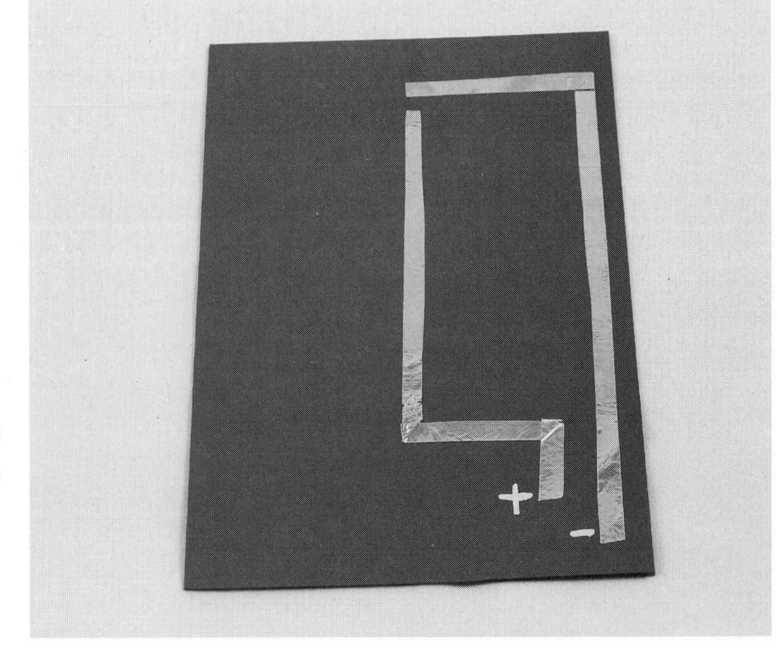

Cut off a piece of copper tape long enough to reach the first LED sticker. Using your design as a guide, peel back one of the edges of the backing of your copper tape and start sticking it down along your positive path. Remove the backing bit by bit to keep your length of tape from sticking to itself.

When you get to a corner, pull the tape in the opposite direction to the path. Flatten the turn, and then pull the tape in the right direction. Flatten this turn too, and then keep sticking your tape to your path until you get to the space where the LED should go.

Tear your copper tape so that it ends in the space for your LED sticker. Then repeat this process for the all other parts of your circuit. Make sure that the positive and negative bits of the copper tape are not touching each other; otherwise, your circuit won't work.

Finishing the Circuit for Your Greeting Card

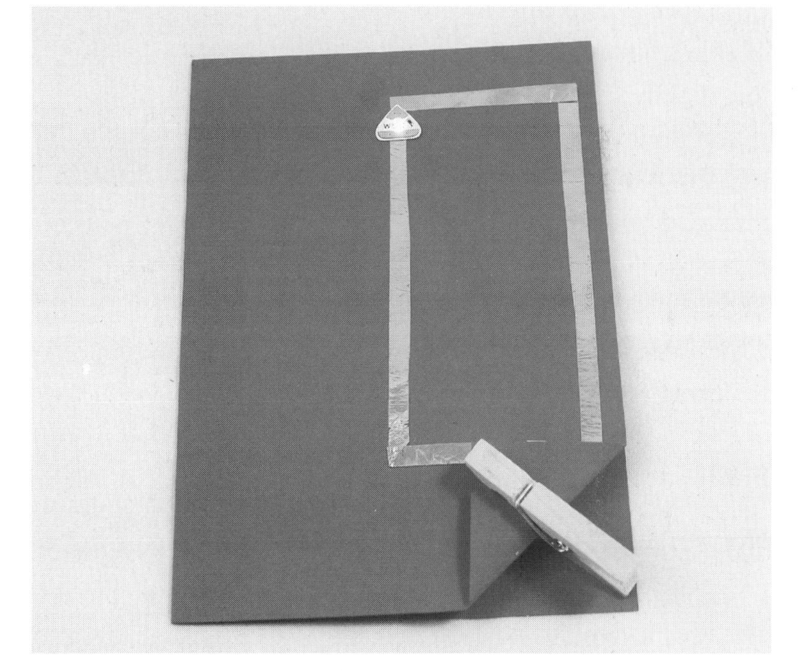

Crease the corner of the card where your battery goes, and place the battery in the corner flap so that the positive side of the battery is connecting to the positive path and the negative side of the battery is connecting to the negative path. Clip the battery in place with a wooden clothespin or a bulldog clip. If you position your clothespin or clip at a right angle, it will help your greeting card stand upright.

Make sure the positive and negative paths are not touching each other. Then stick the LED sticker on top of the copper tape. Pay attention to the way in which you stick your sticker. The positive side of the LED sticker should connect to the positive side of the battery, and the negative side of the LED sticker should connect to the negative side of the battery.

Does it light up? Awesome! You have just made your first unique paper circuit. Is your circuit not working yet? Skip to the end of this project for my troubleshooting guide.

Finishing the Design for Your Greeting Card

The last thing you need to do is add the rest of your greeting card design around the circuit. As you're adding in your design, be careful not to rip your copper tape circuit.

One finishing touch that makes a greeting card look extra special is to cut a sheet of thin white paper so that it is slightly smaller than the size of your card. You can fold this paper in half and then glue it inside the card along the internal back edge. You can use this paper to write a personal message to the person for whom you made the card.

Fix It

Not working? Don't worry! Follow these steps to figure out why, and fix it.

1. Check your power.
   - Is your battery the right way around? Flip it over and see what happens.
   - Has it run out of juice? Try another battery.
   - Is your battery connecting to your circuit? Make sure that both the negative path and the positive path are connected to the correct side of the battery.
2. Check your components.
   - Is your Chibitronics LED sticker the right way around? Remember that the positive side of the battery should be connected to the positive bit of your LED, and the negative side of the battery should be connected to the negative bit of your LED.
   - Is your LED securely stuck in place? Loose connections mean that your circuit won't work.
   - Is your sticker on top of an unbroken bit of copper tape? Electricity likes to take the easiest route possible, so if you put an LED sticker on top of an unbroken bit of copper tape, the electricity will go through the copper tape instead of the sticker.

3. Check your wiring.
   - Do your positive and negative paths touch? If they are touching, no matter how slightly, your circuit won't work.
   - Are all your paths complete? If your copper tape path has rips or tears in it, electricity may not be able to get through.
   - Have you made your path by sticking one bit of copper tape on top of another? Even if your glue is conductive, this can lead to unreliable connections, meaning that your LED might flicker or not work at all.

Make More

Now that you've made your first custom greeting card, why not make more? You could even make a giant collaborative card for a special occasion by teaming up with your friends and family to make one big card with several different circuits on it.

PROJECT 2

Dancing Origami Ladybird

Explore the art of origami and get a wiggle on with vibration motors

Tools

- Sharp scissors
- Ruler, ideally metal

Materials

- Paper, ideally 8-inch square origami paper, red on both sides
- Black paper for decoration
- Regular and double-sided sticky tape
- Glue stick
- Copper tape
- One sewable/stickable vibration motor
- One 3V battery

Origami is the Japanese word for "paper folding." *Ori* means "to fold," and *kami* means "paper." Origami is the creation of paper shapes made entirely by folding. By learning some simple origami techniques, you can transform a simple square of paper into an animal, a fish, a puppet, a toy, or a beautiful geometric shape.

This simple origami-inspired ladybird will show you some of the basic paper crafting skills involved in this ancient art. This make isn't pure origami because we are not just folding the paper; we are adding in a couple of decorations, snipping off a few bits of paper, and of course, making our circuit. I've modified a classic design of an origami ladybird so that it can wiggle and dance without falling over.

Preparing Your Materials

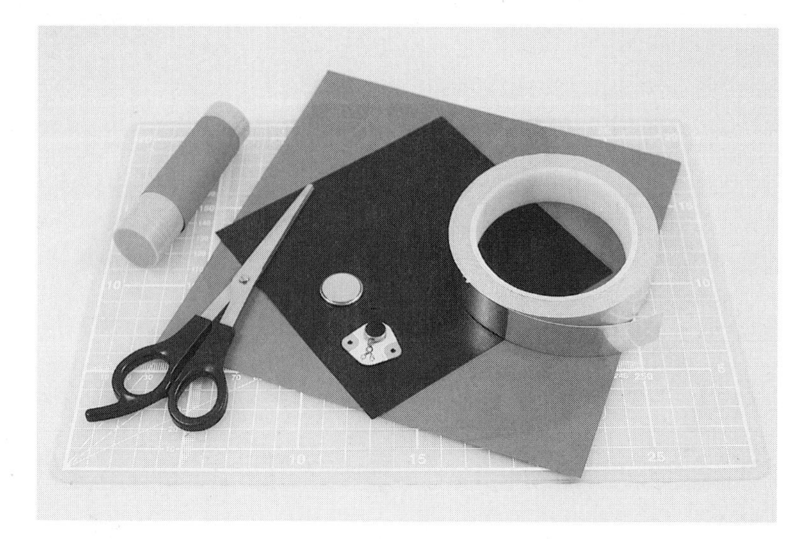

Make sure that you've got all your tools and materials ready. Then prepare your paper. If you are using ordinary red and black paper instead of origami paper, you'll need to fold and trim it into a square first.

To make a square, lay your paper out in landscape mode so that the long edges are at the top and bottom. Fold the top right-hand corner of your page down toward you so that the right-hand short edge of the paper lines up with the long edge at the bottom. Cut off the rectangular bit of paper sticking out from the edge of the folded triangles. Then unfold your paper to reveal your square.

Once you have a piece of square paper, fold it diagonally in half so that the paper makes two triangles; then unfold. Using the crease as a guideline, cut your paper in half. Each of these triangles will make one ladybird. You can make both and give one to a friend, or you could keep it to give your first ladybird some company.

How to Make Your First Origami Folds

Before you start making origami, you should make sure that you are folding on a clean, dry, flat surface. For the neatest folds, align your sides and corners before folding. Then fold firmly. Origami artists use a folding tool called a *bone folder*, but you can crease your folds with your finger, a ruler, or the blunt edge of a butter knife.

The two most basic folds in origami are the *valley fold* and the *mountain fold*. Every origami design starts with one of these two folds. Take two pieces of paper, and try both folds now.

With your first piece of paper, fold it in half by bringing the edge of the paper furthest away from you to meet the edge of the paper closest to you and then creasing. When you open the paper back up again, this fold should look like a valley, sinking down to the crease and back up again.

With your second piece of paper, fold it in half by bringing two edges together underneath the paper and then creasing. When you open it again, this fold should look like a mountain, rising up to the crease and back down again. You can make either fold in any way on your paper—diagonally, on a corner, or in the middle. You may have already noticed that if you flip over a valley fold, you will get a mountain fold, and vice versa.

Making Your First Folds

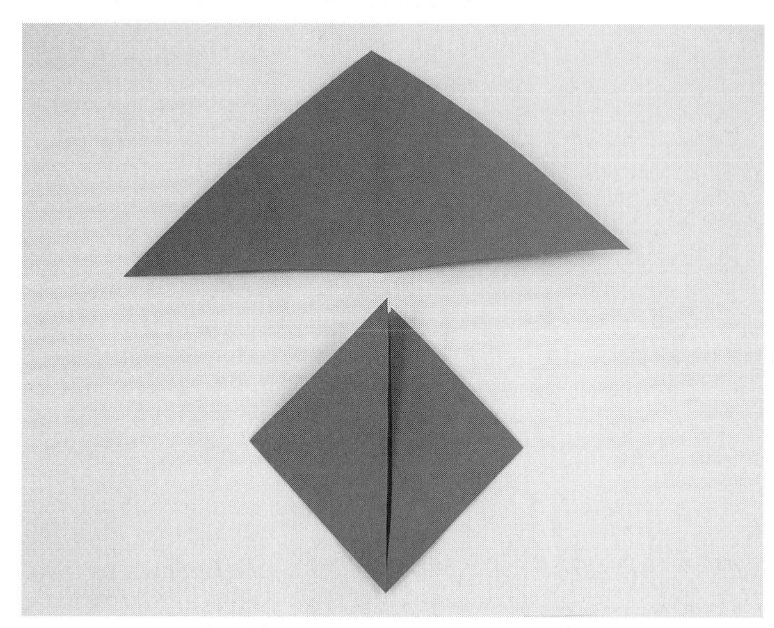

Take one of your red triangles, and place it on a flat surface in front of you, with the long edge of the triangle facing toward you. Fold your triangle in half by pulling the left and right corners together. Align the corners and edges carefully, and then firmly crease the paper along the fold.

Unfold the paper, and you should have a valley fold in the center of your triangle.

Next, fold up the bottom two corners so that the tips meet at the top of the triangle. Once the edges and corners are aligned, firmly crease the paper.

This time, do not unfold the paper. You should now have a diamond shape, just like in the picture.

Folding Your Ladybird's Shell and Head

You now have a closed edge facing you and two flaps at the top of the diamond shape. One at a time, fold down the two flaps from the top until the tips of the flaps are at the bottom of the diamond. Fold the tips down at a slight angle so that they are apart from one another, just as in the picture. These two flaps are going to be your ladybird's shell.

Next, take the back layer of paper that forms the tip of the diamond and fold it toward the center of the paper. Don't fold this bit of paper exactly in half. You need to leave about a quarter-inch gap. Now fold that back layer of paper in half again so that you have a narrow strip of paper with a quarter-inch gap above the central crease.

Flip the whole paper over, and add a strip of thin black paper to the narrow band of folded paper, as in the picture. When folded, this will be your ladybird's head. If you don't have black paper, you can color the strip in with a black felt-tip pen.

Flip the whole paper back over, and get ready to start folding again.

Making Your Final Folds

Fold over the narrow strip along the central crease so that the black part is showing.

Now make a mountain fold on the left and right corners of your ladybird. Fold each of the corners underneath the body of the ladybird until you have a shape that resembles the round body of a ladybird.

Once you have a shape you're happy with, crease the paper firmly. At this point, you are creasing several layers of paper, so you may want to use the blunt edge of a butter knife to help you get a clean, firm fold.

Finally, snip a little of the pointy wingtips off so that they look more like the rounded shell of a ladybird.

Adding the Final Touches

Use your finger to gently poke between the shell and the body of your ladybird, encouraging the wings to come up and the body to fold down a bit. This makes the ladybird a bit more three dimensional but also gives us room to add the electronics that will make this origami ladybird dance!

Now you have to choose how many spots to give your ladybird. There are lots of species of ladybird with different numbers of spots, different markings, and different colors. You can get red ladybirds with two, seven, twenty, or even more black spots. You can get black ladybirds with red spots, yellow ladybirds with black spots, and even striped ladybirds.

I've chosen to make a two-spot ladybird. Whichever type of ladybird you choose to make, you can cut out your spots from another bit of paper and stick them on with glue or simply draw them on with a felt pen.

Preparing Your Circuit

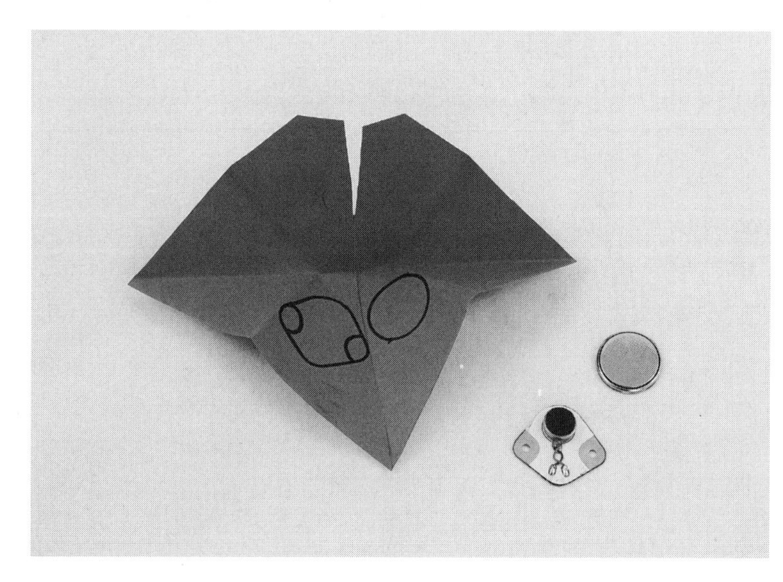

To add in our circuit, we need to unfold our ladybird a little bit. Unfold the left and right corners that are tucked under the bottom of your creature. Then lift up the wingflaps.

We're going to place our components on opposite sides of the body near the head of the ladybird so that it balances nicely and doesn't fall over when it's wiggling.

Put the sewable/stickable vibration motor on the left-hand side of the body and the 3V battery on the right-hand side of the body. Then draw around the components. They should sit on either side of the main bit of the body, underneath the wings when the ladybird is folded back up again.

Once you've drawn around them, take your sewable/stickable vibration motor and your battery off the ladybird and start making your copper tape circuit.

Starting Your Circuit

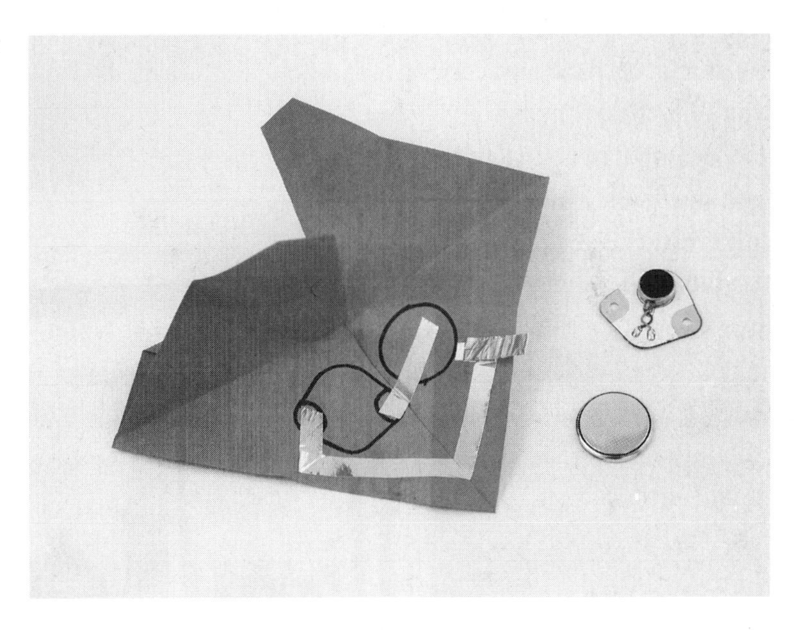

Take a small piece of copper tape, and make a path between the right-hand tab of the vibration motor and the middle of the battery circle, just as in the picture.

Take a length of copper tape, and make a path from the left-hand tab of the vibration motor. Take the copper tape path around the top of the body and then bring it down toward the battery circle. When you are about ¼ inch away from the battery circle, pause and trim off the tape until you have about 1½ inches of tape left. Pull the backing off, and fold the remaining tape in on itself so that the two sticky sides stick together. This should leave you with a flap of double-sided copper tape over the battery circle.

Use double-sided sticky tape to place your vibration motor on top of the copper traces, making sure that your sticky tape doesn't get in the way of the connections. The vibration motor not connecting properly to the copper tape circuit is the most common reason that this make doesn't work the first time. To make a more secure connection, you can put ½ inch of upside-down copper tape on sticky tape (sticky side to sticky side), and then tape it on top of each of the vibration motor tabs.

Finishing Your Origami Ladybird

Put your battery on top of the circle, underneath the flap. Stick the flap to the top of the battery with a bit of sticky tape. You might want to use a little extra tape to secure the whole thing to the body of your ladybird. The bottom bit of tape connects with one side of the battery, and the top flap connects with the other. It doesn't matter which way around you put your battery because the motor doesn't care which direction it gets its power from.

Once your battery is in, your ladybird should start wiggling. Tuck the corners back under the body of the ladybird, and put it on the floor or a table to watch it dance. To stop your ladybird dancing, carefully lift up the battery flap and remove the battery.

Fix It

Not working? Don't worry! Follow these steps to figure out why, and fix it.

1. Check your power.
   - Has your battery run out of juice? Try another battery.
   - Is your battery connecting into your circuit? Make sure that both sides of the battery are connected securely and that the flap on top of the battery is not touching the bottom or sides of the battery. Make sure that your sticky tape isn't interfering with your connections.
2. Check your components.
   - Is your vibration motor securely stuck in place? Loose connections mean that your circuit won't work. The dancing of your ladybird sometimes means that your connections can wiggle loose. Try sticking your motor down with tape or using a bit of upside-down copper tape on sticky tape on top of your vibration motor tabs to make an extra secure connection.
   - Is your vibration motor on top of an unbroken bit of copper tape? Electricity likes to take the easiest route possible, so if you put a component on top of an unbroken bit of copper tape, the electricity will go through the copper tape instead of the component.

3. Check your wiring.

- Do your positive and negative paths touch? If they are touching, no matter how slightly, your circuit won't work.
- Are all your paths complete? If your copper tape path has rips or tears in it, the electricity may not be able to get through.
- Have you made your path by sticking one bit of copper tape on top of another? Even if your glue is conductive, this can lead to a faulty circuit.

Make More

Make several different types of ladybirds with your friends and family, and then have an epic ladybug battle! Draw a circle on the floor, get everyone to place their bugs inside the circle, and any bugs that wiggle their way out of the circle or stop working are out of the game. Which ladybird will be the last bug standing?

PROJECT 3

Cardboard Doorbell

Level up your paper circuits skills by making a doorbell for your bedroom

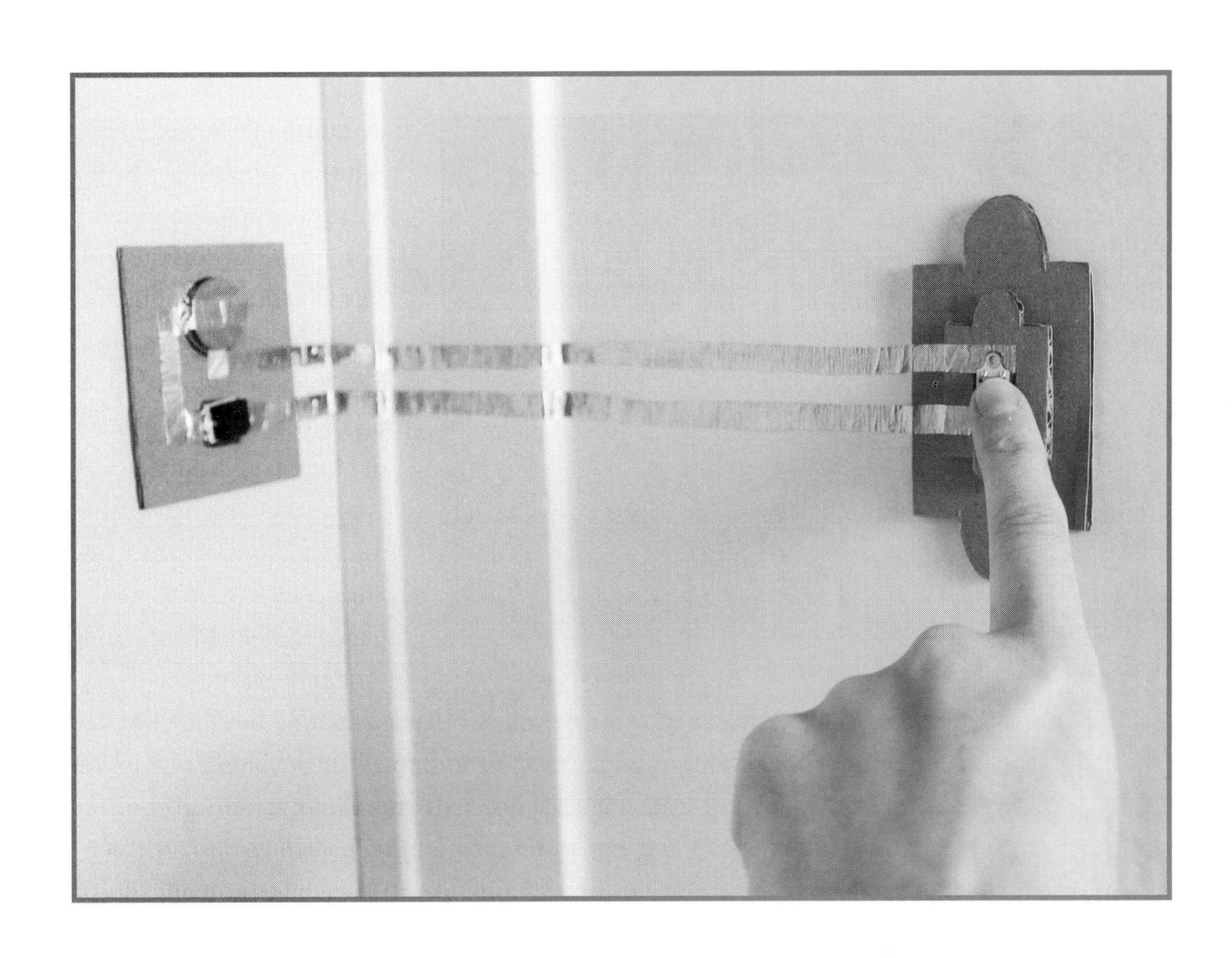

Tools

- Sharp scissors
- Ruler, ideally metal

Materials

- Scrap cardboard
- Copper tape
- Ordinary and double-sided sticky tape
- One sewable/stickable buzzer
- One sewable/stickable push button
- One 3V battery

Copper tape circuits aren't just for making cool pictures on paper. With a bit of creative thought, you can use copper tape circuits in your home or school to light things up, make things buzz, or move things around. Once you get the hang of using copper tape, you can use it around corners, under tables, over doors, or even on furniture to make it do something awesome.

In this project, we're going to learn how to work with cardboard. Cardboard is one of the coolest materials out there. We're going to make a copper tape and cardboard doorbell so that you have a DIY way for visitors to say hello before they come into your room.

If you're putting anything up on a wall, it's polite to ask the person who owns that wall first so that any damage to the wallpaper or paint work is anticipated and acceptable.

Preparing Your Materials

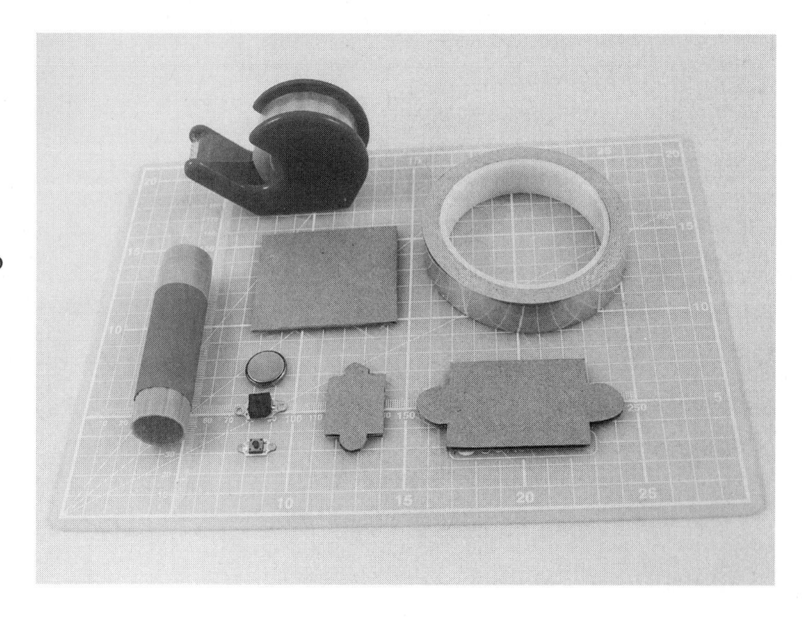

Make sure that you've gotten all your tools and materials ready. Then prepare your cardboard. Go to page 248 for a traceable template for all the parts you'll need to make this cardboard doorbell.

You can mark a template onto paper or card in a

number of different ways. For this project, we'll be keeping it simple by cutting out a paper pattern.

Take some ordinary paper, and place it over the doorbell template on page 248. Using a pencil, trace all the different parts, and then use a pair of scissors to carefully cut them out. Next, place the paper parts onto the cardboard, and secure them in place using some sticky tape or removable masking tape. You can then carefully cut around the shapes before removing the paper.

Don't worry if it takes you a few tries to get your doorbell perfect! I've made a basic shape for the template, but feel free to add your own flourishes.

How to Work with Cardboard

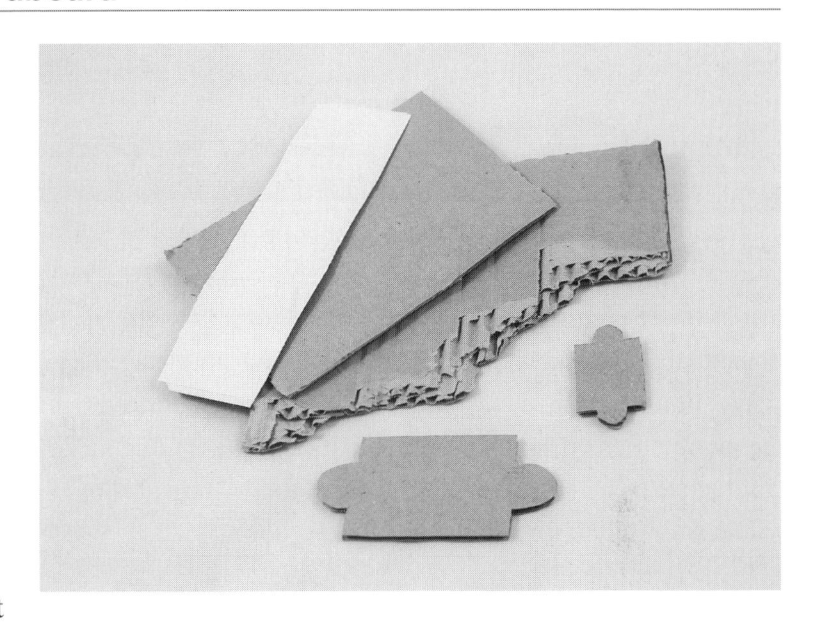

Cardboard is an awesome making and crafting material. You can find it pretty much anywhere for free. Cardboard can be ripped, cut with scissors or craft knives, or even made into elaborate designs with the help of a laser cutter.

Cardboard comes in lots of different types, from the thin stuff that makes your cereal box to the thicker stuff that your deliveries get packaged in. Take a closer look at some of the thicker cardboard around, and you'll see that it has bumpy ridges between flat edges. If it has one layer of ridges, it is called *single walled*, and if it had two layers it is called *double walled*.

You can use whatever cardboard you like in this project, but I'm using a single-walled cardboard just under ¼ inch thick.

If you're making a cardboard project, try using masking tape instead of ordinary sticky tape because sticky tape will peel off layers of the cardboard if you need to move the tape to adjust something. If you want to make a cardboard project with corners, you can fold it by carefully scoring one side of the cardboard with a craft knife (don't go all the way through). Then crease it along the edge of a ruler.

Constructing Your Doorbell Base Plates

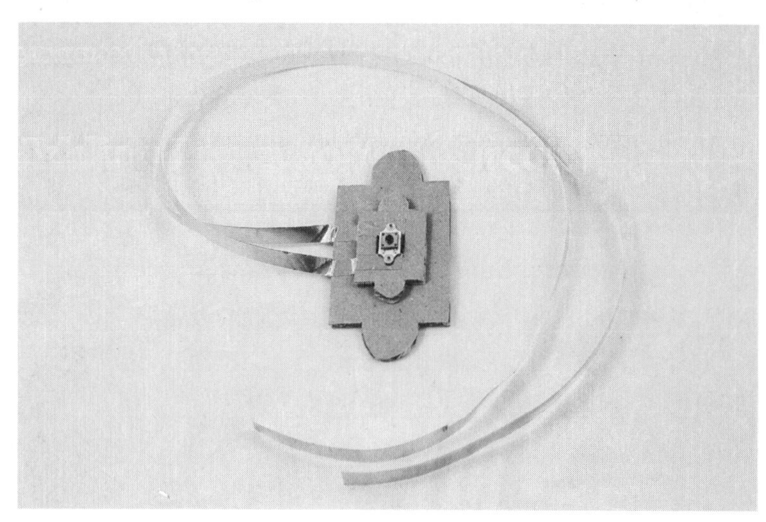

Take the two doorbell base plates you cut out of cardboard earlier, and use double-sided sticky tape to attach the smaller one to the bigger one. Next, take your sewable/stickable button and place it on top of the smaller base plate. Don't stick it down yet, but do take careful notice of where the two metal parts of the button sit because we'll need to make sure that our copper tape paths line up.

Tear off two 15-inch strips of 5-mm copper tape, and peel back the end of one of them ½ inch. One at a time, stick the copper tape onto the smaller base plate in two parallel lines, just as in the picture. Make sure that they do not touch. Once they are in place, press the nonstick backing back onto any exposed tape, and leave the ends loose.

Use double-sided sticky tape to stick the button onto the doorbell base plates, as shown in the picture. Remember that ordinary sticky tape is nonconductive, so don't stick it over the conductive tabs of the button or the conductive tape. Using double-sided sticky tape, put the whole doorbell up on the outside of your door.

What Are Buttons and Switches?

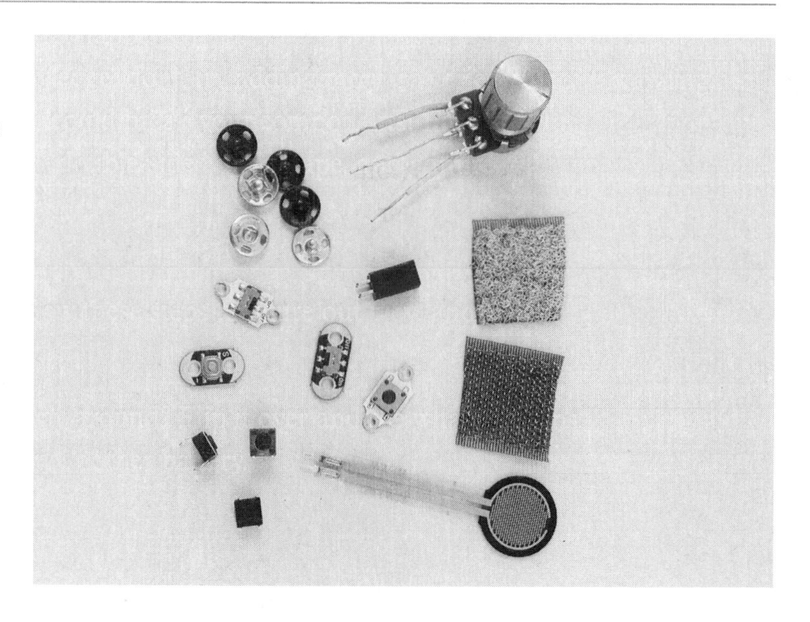

You'll notice that our button does not have a + or – symbol. This is because the button, unlike an LED, which has positive and negative bits, can be used either way around. This means that it doesn't matter which way you connect it into your circuit, as long as the two paths do not cross.

A button is a type of switch. A switch controls

whether a circuit is open or closed. There are many different types of switch that you can choose to use in your inventions, but they all fall into two broad categories: momentary switches and maintained switches.

Momentary switches are only active when they're being pressed, for example, keys on a keyboard or a doorbell. The button we're using in this project is momentary. A *maintained switch* stays in one state until you change it into another, for example, an ordinary light switch that flips between on and off.

In this part, we'll be using both types of switches. Later in this book, we'll make and use other types of switches, including a DIY tilt switch and a DIY pressure-sensor switch.

Planning Where to Put Your Doorbell

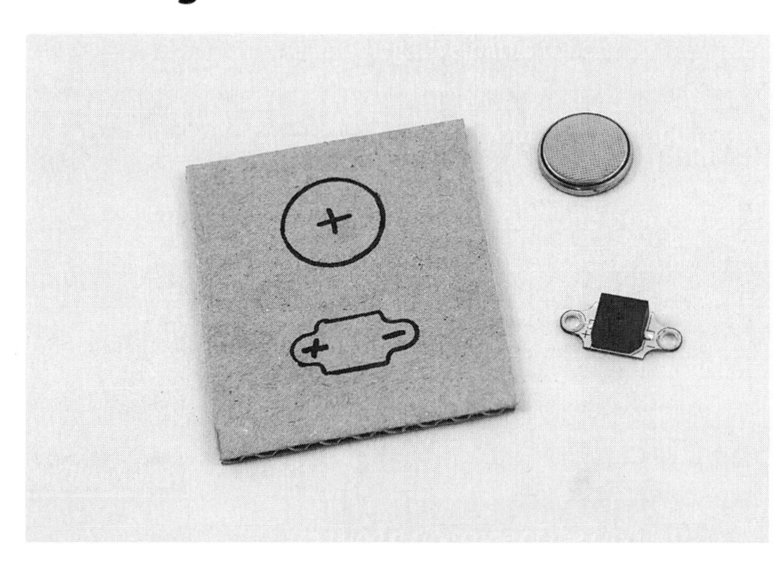

Now that your cardboard doorbell with its button is ready, we need to make the other parts of the project: the buzzer and the battery pack. Think about where you're going to put this other half of your circuit.

I chose to place my buzzer and battery on the inside of my door frame, about 5 inches away from the doorbell. When you're deciding where to place your buzzer and battery, make sure that it's not too far away from your doorbell. Read the following section to find out why.

Once you know where to place it, take out the buzzer base plate you prepared earlier. Place your buzzer at the bottom of the cardboard so that the positive is on the left and the negative is on the right. Draw around your buzzer, and then add symbols so that you know where the + and – are. Place your battery at the top of the cardboard, negative side down, and draw around that too. Using double-sided sticky tape, stick the cardboard in place on the wall, but don't add the components just yet. We need to make our paths first.

How to Move Electricity Along a Path

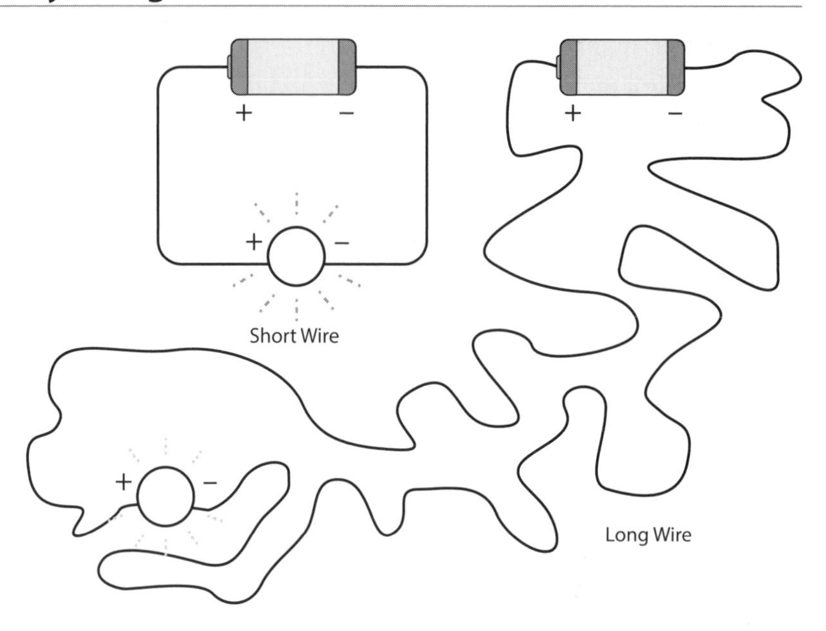

Lengthening your path in a circuit means that the electricity has further to travel. Try to keep your paths to a reasonable length because too much distance could make your LEDs dimmer, your buzzer quieter, or your motors less powerful.

The distance covered on a sheet of paper won't make a big difference, but if you wanted to make your electricity travel all the way across a room, you would see the effect.

Why not try conducting an experiment to see how far away you can get an LED to light up using copper tape as a path?

Making Paths for Your Circuit

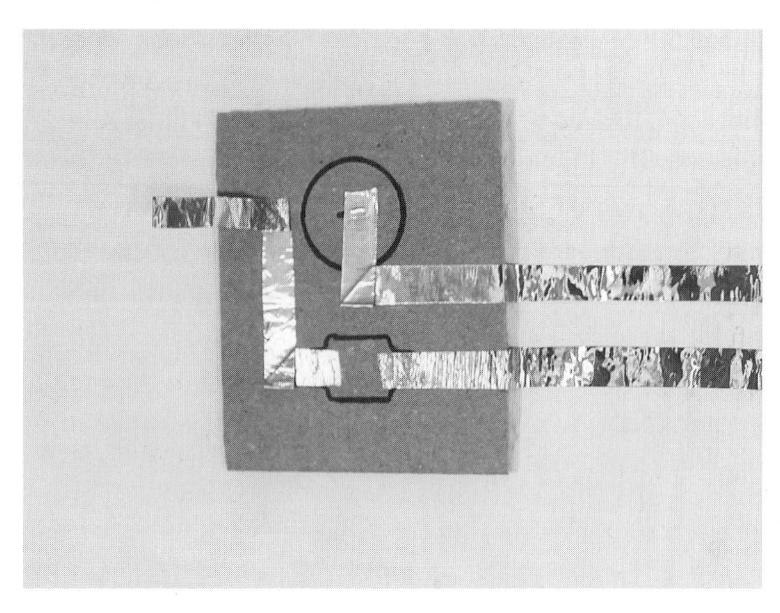

Go back to your copper tape paths attached to the cardboard doorbell. Inch by inch, take the backing off the top path and stick it down, taking the copper tape on a path toward the circle you drew around the battery. When you reach the center of the battery circle, rip or snip off the tape and smooth it down.

Next, take the bottom piece of the copper tape

attached to the cardboard doorbell and take it to the shape you drew around the buzzer. When you get to the space for the negative bit of the buzzer, rip or snip off the tape and smooth it down.

Take a new bit of copper tape, about 5 inches long, and stick one end of it in the space for the positive bit of the buzzer. The two ends of the copper tape should be quite close but not touching. Bring the path in line with the battery circle, and then make a corner with the tape. Continue sticking for about ¼ inch, and then pause. You should have about 2 inches of tape left. Pull the backing off and fold the remaining tape in on itself so that the two sticky sides stick together, making a flap for your battery.

Finishing Your Cardboard Doorbell

Now your circuit is ready for your components. Use double-sided sticky tape to stick the back of your buzzer to the cardboard, over the copper tape. Make sure that you are connecting the metal bits of the buzzer with the copper tape. You may need to secure your buzzer in place with a little more copper and sticky tape later when you test your circuit.

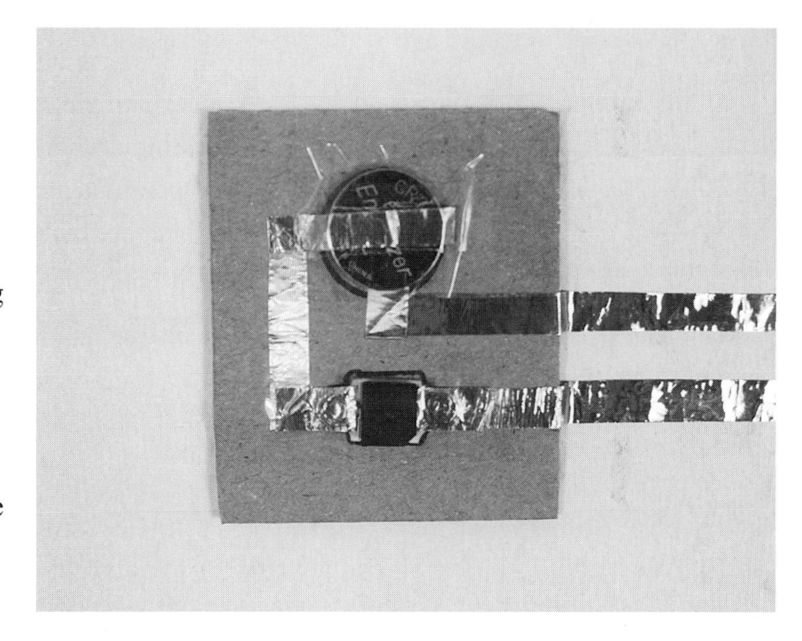

Next, place your battery inside your copper tape flap. Make sure that you place it so that the positive side of the buzzer is connecting with the positive side of the battery via your DIY copper tape flap. The negative side of the battery should connect with the negative side of the buzzer via the button. You should also be careful that the top bit of the copper tape flap is not touching the bottom path, or your circuit won't work. Secure the battery in place with tape, and then test your circuit by pressing the button.

Testing Your Cardboard Doorbell

If your cardboard doorbell doesn't work on the first go, it's time to troubleshoot! At the end of each project, I've added tips and tricks to help you figure out what to do. I've been making things for many years, and even now I don't always get things right the first time. One of the most important skills you learn as a maker is to keep trying.

In this project, the most common problems are that your battery or buzzer is the wrong way around or that your connections are not secure enough. Try gently pushing down the positive and negative sides of the buzzer while you're pressing the button—does it work now? If so, you know that it is a problem with the connection.

The buzzer needs a connection to work, so you can make it more secure by adding copper tape on top. Take a bit of copper tape and stick the sticky side to the sticky side of regular sticky tape, just as in the picture. You can use this sticky tape to connect the top of your buzzer tabs to your existing copper tape paths.

Fix It

Not working? Don't worry! Follow these steps to figure out why, and fix it.

1. Check your power.
   - Is your battery the right way around? Flip it over and see what happens.
   - Has it run out of juice? Try another battery.
   - Is your battery connecting into your circuit? Make sure that both the negative path and the positive path are connected to the correct side of the battery and that the positive and negative paths are not touching.
2. Check your components.
   - Are all your components the right way around? Remember that the positive side of the battery should be connected to the positive bit of your buzzer, and the negative side of the battery should be connected to the negative bit of your buzzer via your button.

- Are all your components securely stuck in place? Loose connections mean that your circuit won't work.
- Are any of your components on top of an unbroken bit of copper tape? Electricity likes to take the easiest route possible, so if you put a component on top of an unbroken bit of copper tape, the electricity will go through the copper tape instead of the component.

3. Check your wiring.
   - Do your positive and negative paths touch? If they are touching, no matter how slightly, your circuit won't work.
   - Are all your paths complete? If your copper tape path has rips or tears in it, the electricity may not be able to get through.
   - Have you made your path by sticking one bit of copper tape on top of another? Even if your glue is nonconductive, this can lead to a faulty circuit.

Make More

Why not try adding a sewable/stickable LED to your circuit for a visual effect. You can also add an on/off switch so that you can choose not to be disturbed.

PROJECT 4

Flickering Firefly Wall Art

Make your own DIY switch in this interactive wall art project

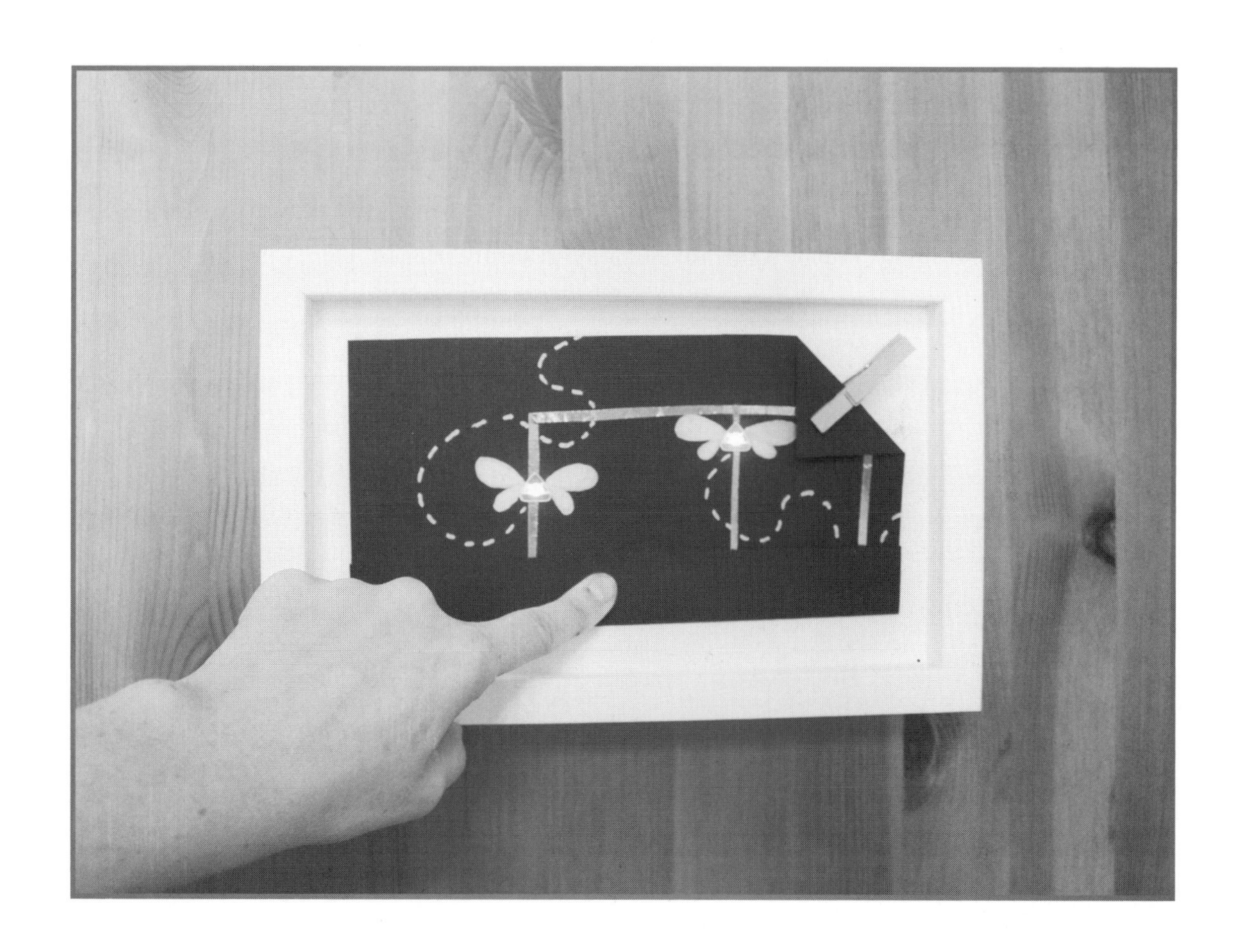

Tools

- Sharp scissors
- Ruler, ideally metal

Materials

- One piece of navy or black A4 card
- One piece of thick white A4 card
- White photo frame
- Copper tape
- Marker with silver ink
- Two Chibitronics LED stickers
- One 3V battery and a wooden clothespin or small bulldog clip

One of my favorite insects is the firefly. However, I grew up in a cold, rainy country called Wales where fireflies don't generally like to hang out. I was a very excited adult when I saw my first firefly, much to the amusement of all my friends! Here's my firefly-inspired interactive wall art using one of my favorite paper circuit techniques: the flicker switch.

The first DIY paper circuit switch I ever made was from a template designed by Jie Qi. As you know, she invented the Chibitronics LED stickers that we're using in this book. Of all of Jie Qi's designs, the one that really captured my imagination was her flicker switch. Since I learned about it, I've used it in many different projects. This project is a remix of her original flicker switch adapted to add in an extra LED sticker using our first parallel circuit.

Preparing the Frame

Make sure that you've gotten all your tools and materials ready. Then prepare your card and frame. Open up your photo frame, and pull out the glass or transparent plastic at the front. Your art is made to be touched, so we are not going to be using the glass or plastic.

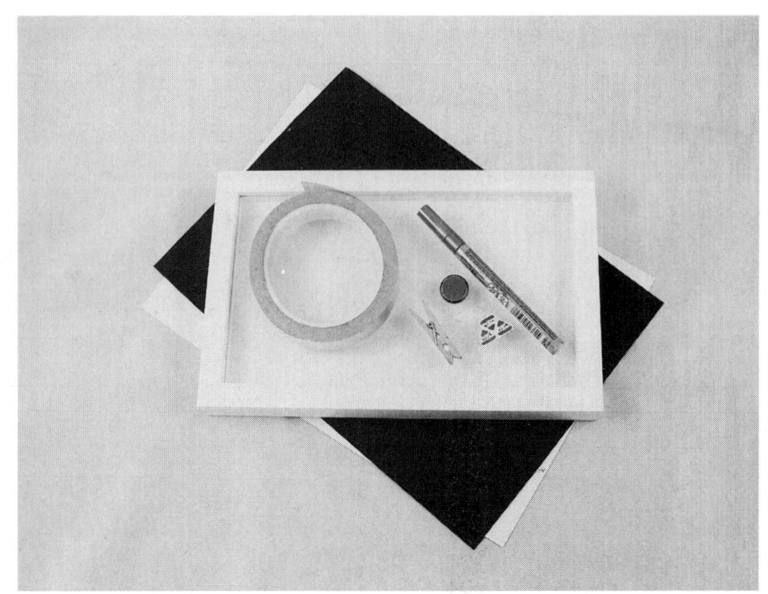

You should now have the back of the frame, the frame, and the card or paper insert that was between the back of the frame and the glass. You need to cut your white card to fit exactly where the glass went. The easiest way to do this is to carefully use a craft knife to cut around the glass, but if you're not confident doing this, you can use the paper insert to trace an outline, and then cut the shape with scissors.

Make sure that your white background card fits securely and will not fall out of the frame. You may need to add extra cardboard "padding" behind the card to keep it securely in place.

Preparing Your Card

Next, you need to cut out the navy or black card. It should be smaller than your white card so that you can see the edges of the white card inside the frame. You'll need to allow an extra bit of room at the bottom for a flap that will become the flicker switch. I'd recommend testing the sizing and folding with a bit of scrap A4 paper before cutting your card.

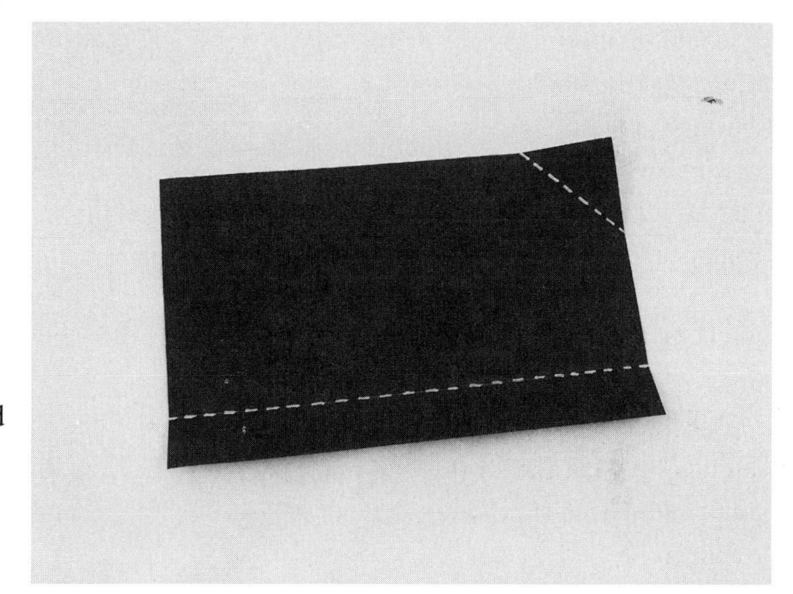

Once you've cut your card to the right size, you'll need to make two folds. The first is for your 3V battery, in the top right-hand corner. Make sure that your battery fits first, and then fold over the corner and give it a good firm press to keep it in place. I've marked the fold positions in the picture.

Finally, fold the long flap at the bottom of your card as shown in the picture. You are now ready to start building your circuit.

If your navy card is thick and you want to get an extra precise fold, you can try a technique called *scoring*. The next section tells about this useful craft technique.

How to Score Card

If you want to get an extra precise fold, you can try a technique called *scoring*. You'll need a ruler and a butter knife (easy version) or a craft knife (expert level).

For the easy (butter knife) version, line your card up with a ruler wherever you want to make the inside of your fold. Hold your ruler firmly, and use the back of your butter knife (the nonserrated edge) to firmly push down along the length of your ruler in a sliding motion. Your card should now be firmly indented, making it easier to fold in a neat line.

For the more difficult (craft knife) version, flip your card over so that the outside of your fold is showing. Line up your ruler, and use your craft knife to very carefully make a light cut along the length of your fold.

You should only cut the card slightly, not go through the whole thickness of the card. Be careful not to cut too far through your material. When you flip the card back over and fold it in, the outside of the fold will now have more "give," meaning that you'll get a crisper-looking fold.

Starting Your Flickering Firefly Circuit

Prepare your copper tape. For our previous projects we have used 5-mm copper tape, but because this project looks better with thinner lines, I used scissors to cut the tape in half. If this is too fiddly for you, just use standard 5-mm copper tape. It will still look great on your wall. You'll need to prepare two or three 12-inch lengths of thin copper tape for this project.

Start applying your tape to the circuit. Do your positive path first, starting from the battery circle and heading down to the fold. Make a neat left turn just before the fold (if you can't remember how to make one, take a look back at "How to Work with Copper Tape" on page 10, and continue taping until you're near the end of the card. Rip or snip off the end of your copper tape.

Next, we are going to make our first parallel circuit, which means that we can use two LED stickers in this project.

How to Make a Parallel Circuit

As we found out earlier, there are two different ways to connect up more than one component. One of the ways we can make a project with more than one LED is to connect them in a parallel circuit.

To make a parallel circuit, we use one path to connect all the positive bits of our components (in this case our LEDs) to the positive bit of the battery. We then use another path to connect all the negative bits of our components to the negative bit of the battery.

In a parallel circuit, if one component is faulty, the other parts of the circuit will still work. Another cool thing about a parallel circuit using LEDs is that all the lights will shine brightly, just as if we used only one LED. However, this means that our battery won't last as long because we're using up its power more quickly.

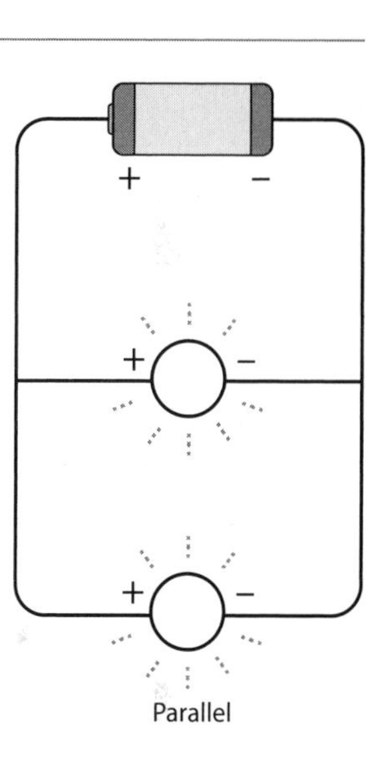

Parallel

Making Your Parallel Circuit

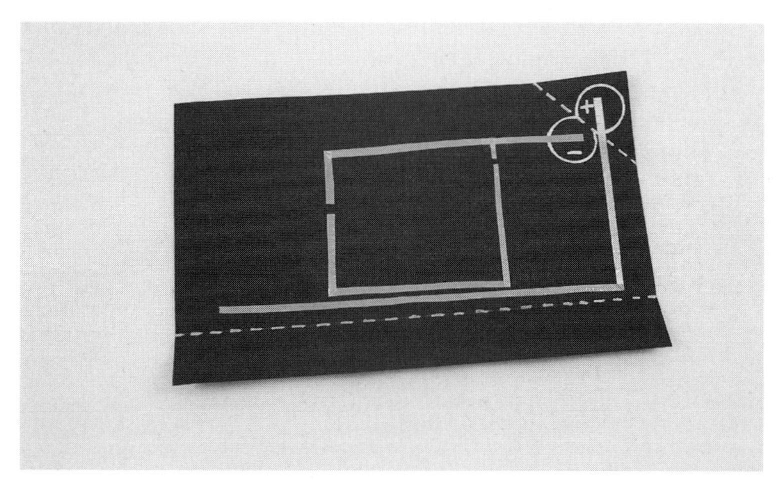

Take another length of thin copper tape and start applying it from the middle of the negative battery circle. Carefully apply it all the way to where the top of your furthest-away bug is going to be. Rip or snip off the end of the copper tape.

Take another bit of copper tape and apply it

close to (but not touching) where you ended the last strip. This is where you are going to put your first LED sticker. Remember that the sticker needs to go on top of the copper tape to work.

Apply a second bit of copper tape in a path to the bottom of the place where your second bug is going to be. Rip or snip off the end of the copper tape. This will join the positive bits of the stickers together, so now we just need to join the negative bits as well by adding a new branch of the path coming from the positive side of the battery.

To do this, take a small bit of copper tape about double the length you need to connect the top of the second LED sticker bug. Fold the tape in on itself so that the two sticky sides stick together. Use a tiny bit of clear sticky tape to stick this down to your card on top of your original path, making sure that you leave enough exposed copper to connect the top of the sticker. Your circuit should look like the picture.

Finishing Your Flickering Firefly Circuit

Take your LED stickers and place them on top of your copper tape. The pointy sides of the stickers are negative, so they should connect with the path coming from the negative side of the battery. The flat sides of the stickers are positive, so they should connect with the positive path of the circuit.

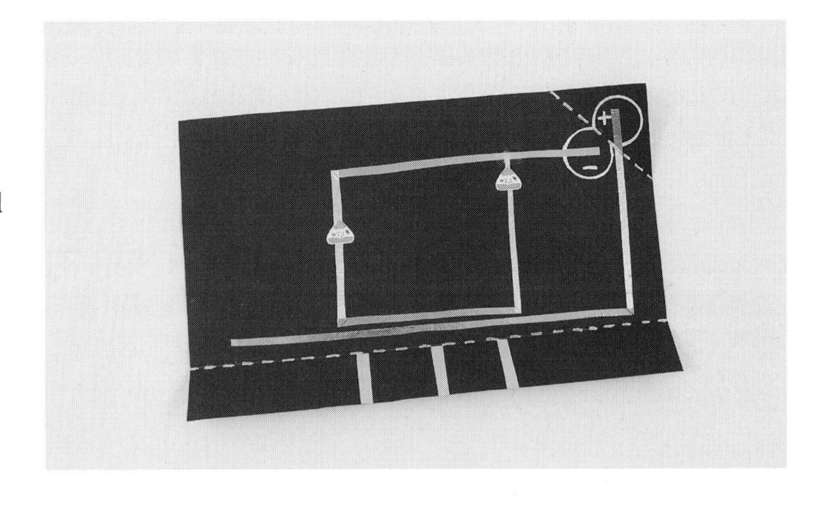

But wait! You may have noticed that the positive path doesn't actually connect with anything yet. This is where the flicker switch comes in.

Take three or four small bits of tape and stick them on the opposite side of the flap to the rest of the circuit, just as in the picture. These bits of tape should align with the part of the circuit where the positive and negative paths run parallel to each other.

When you fold over the flap and run your finger over the hidden bits of tape, they complete the connection in a flickery way as your finger moves. Once you've finished your flicker switch, add in a battery to test your circuit before making the finishing touches to your flickering firefly interactive wall art.

Decorate Your Flickering Firefly Circuit

Now that your circuit is completed, it's time to finish the art. I used a beautiful metallic silver marker to draw in the wings and flight paths of my fireflies. Before you put your designs on your finished circuit, practice getting the effect you want on some scrap card first.

I've chosen a fairly simple decoration for my project, but you could choose to add a tinfoil moon, bright white stars, and some glittery accents or you may glue on some lace or mesh wings to finish off your flickering firefly wall art.

If you are using glitter, do make sure that you read the label. Most glitter these days is made from shiny plastic, but some specialty kinds of glitter (mainly the type of glitter used for makeup) are made with metal such as aluminum. If so, it may be conductive, meaning that the glitter could break your circuits! Tinfoil is also conductive, so make sure any foil stays well away from your circuit.

Frame Your Flickering Firefly Circuit

Circuit working? Art looking gorgeous? It's time to frame your DIY electronics masterpiece. Put your white backing card securely inside the frame, and put the back on. The white card should not wiggle around when you touch it. If it does, add layers of paper or cardboard until it is firmly in place.

Flip your art over and apply glue to the back of the card, everywhere except for about a 1-inch margin around the clothespin that holds your battery in place. Carefully place your art onto the white background in the frame, and firmly press down. Keep the clothespin in place: when your battery runs out, you can carefully open the clothespin and replace the old battery with a new one. Finally, hang your art at touching height, and invite others to enjoy your creation.

Fix It

Not working? Don't worry! Follow these steps to figure out why, and fix it.

1. Check your power.
   - Is your battery the right way around? Flip it over and see what happens.
   - Has it run out of juice? Try another battery.
   - Is your battery connecting into your circuit? Make sure that both the negative path and the positive path are connected to the correct side of the battery and that the positive and negative paths are not touching. If you're using a clothespin to hold your battery in place, make sure that it's pushing on all the way, not connecting at an angle.
2. Check your components.
   - Are all your components the right way around? Remember that the negative side of the battery should be connected to the negative bits of your LEDs and the positive side of the battery should be connected to the positive bits of your LEDs via your flicker switch.
   - Are all your components securely stuck in place? Loose connections mean that your circuit won't work.
   - Are any of your components on top of an unbroken bit of copper tape? Electricity likes to take the easiest route possible, so if you put a component on top of an unbroken bit of copper tape, the electricity will go through the copper tape instead of the component.
3. Check your wiring.
   - Do your positive and negative paths touch? If they are touching, no matter how slightly, your circuit won't work.
   - Are all your paths complete? If your copper tape path has rips or tears in it, electricity may not be able to get through.
   - Have you made your path by sticking one bit of copper tape on top of another? Even if your glue is conductive, this can lead to a faulty circuit.

Make More

Like the flickering effect in this project? Try making a really really, really, long flicker switch with lots of different circuits that tell a story as you move your finger along the wall from left to right. Take inspiration from some of the famous tapestries that told stories to people walking by, for example, the 70-meter-long Bayeux Tapestry that tells the story of the Battle of Hastings in 1066. Epic!

PROJECT 5

Spy Bird

Make a beautiful origami bird that flashes secret messages when you flap its wings

Tools

- Sharp scissors
- Needle-nose pliers

Materials

- Paper, ideally 8-inch-square origami paper
- Sticky tape
- Copper tape
- One 3-mm LED
- One 3V battery

When our spy bird's wings touch, the hidden LED in the folds of its head lights up so that we can silently signal messages. Use spy bird with an existing secret code such as Morse code to send secret messages to your friends. You can also make up your own flashing-light language to share your thoughts with your best friend without everyone else knowing what you're saying.

In this project, we'll be boosting our origami skills by learning a new fold, the *reverse fold.* The first time you do this fold it can seem difficult, but after a couple of tries, it becomes super simple. This project has four reverse folds, so you'll be a pro at using them by the time you've finished Spy Bird! We'll also be using ordinary LEDs instead of stickers for the first time.

Preparing Your Materials

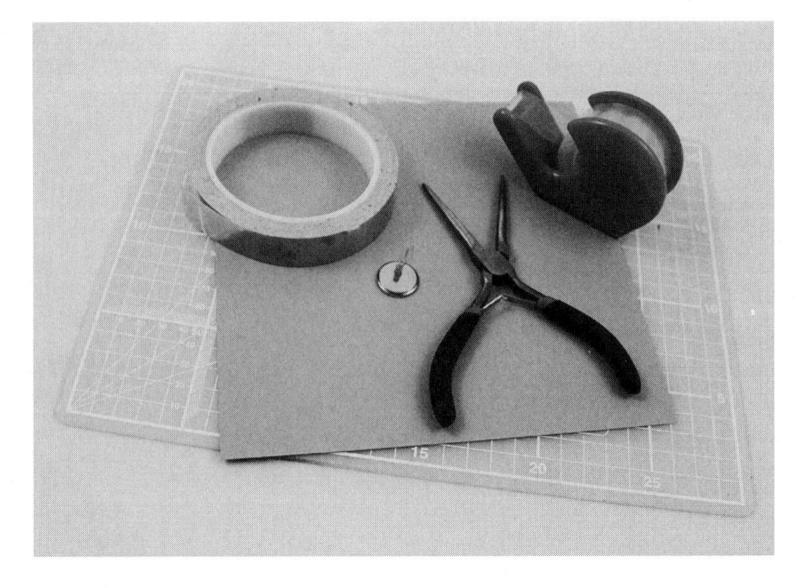

Make sure that you've got all your tools and materials ready. Then prepare your paper for folding.

If you are using ordinary paper instead of origami paper, you'll need to fold and trim it into a square first. To do this, lay your paper out landscape style so that the long edges are at the top and bottom. Fold the top right-hand corner of your page down toward you so that the right-hand short edge of the paper lines up with the long edge at the bottom. Neatly cut off the rectangular bit of paper sticking out from the edge of the folded triangles, and then unfold your paper to reveal your square.

Put your square of paper in front of you at an angle so that it looks like a diamond. Take the bottom corner and fold it to the top corner so that the crease runs from the left corner to the right corner. Make a sharp crease. You're now ready to start folding your spy bird.

Folding Your Spy Bird

You should now have a triangle with the longest edge nearest you. Fold this triangle in half by taking the left corner and folding it to the right. Make a sharp crease.

You now have a triangle with two short edges and one long edge. The short edge to the left should be closed, and the short edge at the bottom should be open, like two flaps. Take the corner of the top flap, and fold it all the way up to the top corner. Make a sharp crease.

Take the point of the new triangle you just made, and fold it to the left-hand edge. Make an extra sharp crease, and then unfold the last fold. Look at the picture to see what you should be aiming for.

For our next step, we're going to learn a new origami fold called a *reverse fold.*

How to Make a Reverse Fold

There are lots of different types of fold in origami. We've already learned how to do a valley fold and a mountain fold, and now we're going to learn another important fold, the reverse fold.

To make a reverse fold, you have to fold and crease the paper and then unfold it again. You then refold and crease the paper using the same crease line but reversing one of the outer creases and the center crease so that it tucks inside itself.

Take the last fold you made and unfolded on your spy bird. Instead of pulling the point over toward the edge of your triangle, open up the flap of paper and push the point inside, along the same crease line.

If you have done your first reverse fold correctly, you should have an upside down triangle tucked inside two layers of a trapezoid. Look at the picture to check that you're on the right track. The folded paper in the lower-left corner is what you should end up with. The bit of paper in the upper-right corner is what your reverse fold should look like on the inside.

Folding Your Spy Bird's Head

Flip your spy bird over and repeat all the steps you just completed on the other side. The reverse fold gets a lot easier the second time you do it!

If you have done your reverse folds correctly, you should have two upside-down triangles tucked inside your paper.

You now have a trapezoid with four sides: two longer parallel sides and two shorter sides. One of the shorter sides has lots of open flaps, and one of the shorter sides has two closed flaps.

We're going to make another reverse fold on the side with the two closed flaps. This will make our bird's head. Put the trapezoid in front of you with the longest edge near you and the short edge with the closed flaps to your right. Take the point at the bottom right and fold it up so that you make a small triangle as in the picture. Make a sharp crease and unfold.

Reverse fold this new little triangle into the center of your folded paper. It should look like a bird's head, as you can see in the picture.

Folding Your Spy Bird's Wings and Tail

Take the top and bottom flaps at the left of your spy bird and fold them up. These are your wings. Your spy bird is starting to take shape. We just need to make one more reverse fold, and then we can start making the circuit.

Take the tail end of the spy bird and fold the point down along the edge created by the body of the bird. Make an extra sharp crease, and then unfold the tail fold. Open up the underneath of the body a little, and push the tail inside the body using a reverse fold so that it looks like the picture.

You're done with the folding! I think it looks a little bit like a hummingbird in flight. Which bird do you think it looks most like?

Our next step is the circuit, but first we're going to prepare an ordinary LED for use in this paper circuit project.

How to Use an LED in a Paper Circuit

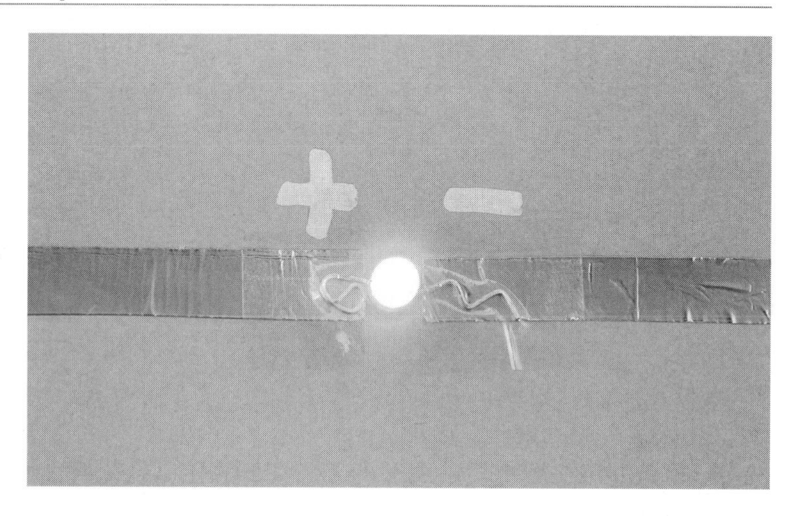

Until now, we've used LED stickers, which are super simple and reliable. Now I want to introduce you to ordinary LEDs. The letters *LED* stand for *light-emitting diode*. Like all diodes, LEDs only let electricity flow in one direction. This is why the LED stickers have positive and negative sides, unlike the vibration motor and push button, which don't mind which way around you put them.

Ordinary LEDs don't have + or – signs, but they do have two different legs that tell us which way around to put them. The longer leg is the positive side, and the shorter leg is the negative side. Here's how to prepare an ordinary LED for use on paper.

Using needle-nose pliers, carefully bend the shorter leg of the LED flat. Then use the pliers to make the wire into a zigzag shape. Be careful not to break the wire by bending the same joint backward and forward too many times.

Next, use your needle-nose pliers to curl the longer leg of the LED into a spiral. Twist this spiral slightly so that it will lie flat on your paper. When you're done, the LED should look like the picture.

Planning Your Negative Path

When your spy bird's wings touch, you want the LED to light up so that you can silently signal messages using a secret code. To do this, you need the circuit to be completed only when the wings touch.

We're going to put our LED inside the fold of the head, on the left. Use a pencil to mark where the positive (front, curly) and negative (back, zigzag) legs of the LED are going to go. Trace from the negative leg of the LED up along the inside edge of the left wing, right to the top. Then trace down from the top inside edge of the right wing down to the bottom. When these two edges touch and separate, they will act like a momentary switch, turning the LED on and off.

Unfold the bird to carry on drawing the negative path. Don't worry—now that the bird has been folded, it will be very easy to fold it back up again.

Make a turn with your pencil, and trace straight down toward the edge. Turn left and finish the path in the middle of a circle, just as in the picture. This will be where the battery goes.

Planning Your Positive Path

Start the positive path with a battery circle. Then take your path toward the central crease of the bird. When you get close to the central crease, turn and take your path to the edge of the paper. When your bird is folded, this edge will connect with the underside of your bird's body, which is where we want our path to go.

To help us draw this bit of the path, fold your square back into a triangle. Now you can take your positive path over the end of the tail and across the underside of the body to reach the front of the head. The sketches of your paths should look like the picture.

Check that the paths you've drawn make sense by tracing the positive and negative paths with your finger. The positive side of the battery should be connected to the positive bit of your LED, and the negative side of the battery should be connected to the negative bit of your LED via your wing-tip switch.

When you're happy that your planned path makes sense, you can start making it!

Finishing Your Spy Bird

Use copper tape to follow the paths you planned, starting from where the negative side of the LED will go and heading up the left wing before stopping. Start again at the top of the right wing, and then follow your path to the negative side of the battery.

Fold your opened square back into a triangle, and then you can take your positive path from the battery, over the end of the tail, and across the underside of the body to reach the front of the head, where your LED will go. Make sure that you leave a small gap between the positive and negative paths where the LED will sit.

Stick the LED over the positive and negative paths with sticky tape. The legs of the LED should be flat against the copper tape. Refold your bird, and then slot your 3V battery into its body.

Hold your bird by the bit of the body with the battery inside, and touch the wings together. Working? Hooray! Use Morse code to send secret messages to your friends, or make up your own light language to share your signals.

Fix It

Not working? Don't worry! Follow these steps to figure out why, and fix it.

1. Check your power.
   - Is your battery the right way around? Flip it over and see what happens.
   - Has it run out of juice? Try another battery.
   - Is your battery connecting into your circuit? Make sure that both the negative path and the positive path are connected to the correct side of the battery and that the positive and negative paths are not touching.
2. Check your components.

- Is your LED the right way around? Remember that the positive side of the battery should be connected to the positive bit of your LED, and the negative side of the battery should be connected to the negative bit of your LED via your wing-tip switch.
- Are all your components securely stuck in place? Loose connections mean that your circuit won't work.
- Are any of your components on top of an unbroken bit of copper tape? Electricity likes to take the easiest route possible, so if you put a component on top of an unbroken bit of copper tape, the electricity will go through the copper tape instead of the component.

3. Check your wiring.
   - Do your positive and negative paths touch? If they are touching, your circuit won't work.
   - Are all your paths complete? If your copper tape path has rips or tears in it, the electricity may not be able to get through.
   - Have you made your path by sticking one bit of copper tape on top of another? Even if your glue is conductive, this can lead to a faulty circuit.

Make More

This was a fairly complex project, both in terms of craft skills and in terms of circuit building skills. If this project has left you wanting more, I think you're ready to tackle a classic origami make: the crane. Look it up online, and think of a way you could make it interactive with lights, a buzzer, or a vibration motor.

PROJECT 6

Pop-Up Cityscape

Combine all your paper circuit skills for this challenging, beautiful make

Tools

- Sharp scissors
- Sharp craft knife
- Cutting mat
- Metal ruler

Materials

- Two pieces of black A4 card
- One piece of white A4 paper
- One piece of tracing paper
- Copper tape
- Glue stick
- Three Chibitronics LED stickers
- Two 3V batteries with a clothespin or small bulldog clip

My favorite time to walk is at dusk. I love looking at the silhouettes of buildings and watching the lights flicker on at night. One evening I realized that these buildings would make a really pretty pop-up project. After some thought, I decided to make a nighttime cityscape combining four buildings in the city of London, where I lived for 10 years before moving to my current home in Berlin.

Pop-up cards look super special, but they are actually very simple to make. This project will also challenge you to practice your free-hand craft knife skills and level up your paper circuitry skills.

Preparing Your Materials

Make sure that you've got all your tools and materials ready. Then prepare your card. If you don't have a good craft knife and a metal ruler, it's worth investing in them for this project. Read my advice on how to use

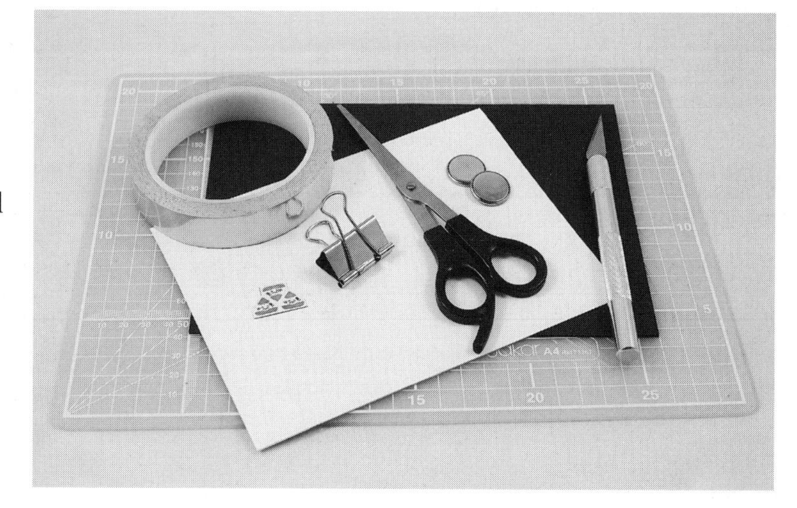

a craft knife on the next page, and remember that sharp knives are much safer than blunt knives.

Fold one piece of black A4 card in half, and then crease it firmly. Next, fold your piece of white A4 paper in half, and then cut off a ½-inch margin from all open sides. This means that your white paper will fit nicely inside the black card, creating a border effect just as in the picture.

If you can, cut your copper tape into thin strips about ⅛ inch wide. If this is too thin for you, strips of about ¼ inch wide will work fine too. You'll need about three 12-inch lengths of copper tape for this project.

How to Make a Pop-Up Card

Start with a simple piece of paper, and fold it in half. Then cut a pair of parallel lines about 1 inch apart into the middle fold. These cuts should be about a quarter of the length of the card. Open up your paper until the two halves are at right angles, and push the 1-inch-long flap through so that it is sticking out in the middle. Refold your paper with your pop-up flap on the inside, and give the folds of the pop-up flap a firm crease. Open your paper up again, and glue a simple shape to one of the sides of the pop-up flap. When you close the paper, the flap and the shape will fold down, and when you open the paper, the flap and the shape will pop up.

That's all there is to a basic pop-up card. You can mix this basic idea up by making different pairs of parallel lines with different lengths, giving a sense of perspective. These cuts should be a maximum of half the length of the card or they'll poke out when you close it. You can also make a pop-up card with several pairs of parallel lines of the same length. This adds stability.

Once you get the hang of this technique, you can do all sorts of interesting pop-up crafts. You can learn to cut pop-up shapes, letters, or even whole words to form intricate and cool designs.

Designing Your Cityscape

Take a piece of scrap paper, and start sketching some buildings you'd like to see in your cityscape. They can be real buildings from your town or a city you'd like to visit, or they can be made-up buildings.

When designing your city, you'll need to add a base for the circuit and a flap to attach it to the pop-up card. The base plus flap will need to be about 1 inch in total. Make your buildings no higher than 3 inches above the base or they will poke out of your card when it is closed.

Try to keep your design simple. Straight lines are much easier to cut than curved lines, and lots of intricate cutouts can take a lot of time to do and make the structure weak. Once you have your outline drawn, sketch out some "windows" in the buildings for the light to come through.

Now that you know what your design is going to look like, cut the pop-up page. I made two pop-up supports, each 1 inch deep. Pay attention to where the supports go on your design—you don't want them appearing in the middle of buildings!

How to Use a Craft Knife

Learning to use a craft knife safely is a very useful skill. However, a knife is a dangerous tool, so you should always be careful and thoughtful when you are using one. Here are the most important things to remember.

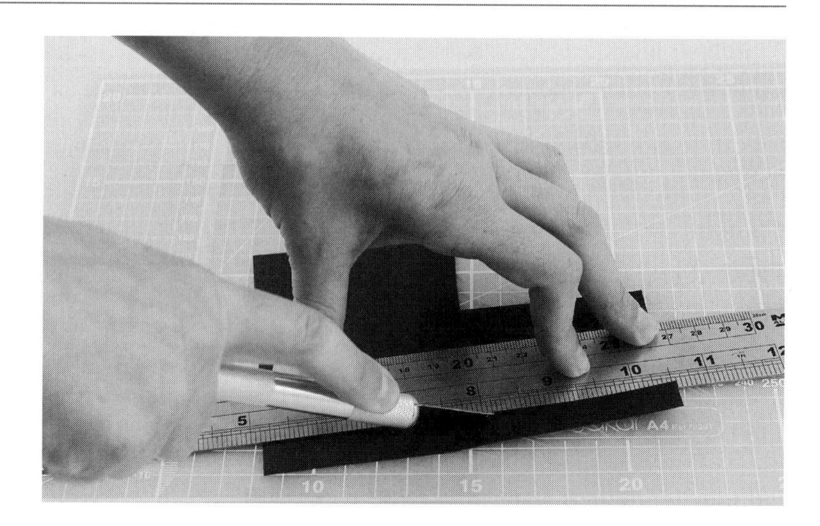

1. Always use a sharp blade. You are far more likely to cut yourself with a blunt knife because the knife is harder to control and more prone to jumping off the paper and onto you!
2. Watch your hand position. Never put your thumb or fingers in the path of the blade.
3. Always keep your hand firmly on the ruler, paper, or cutting mat away from the blade.
4. Use a metal ruler. A blade will cut through plastic rulers, meaning that you will get wonky lines and be more likely to slip and cut yourself.
5. Protect your work with your ruler. Your work should stay under the ruler so that if your knife goes off track, you'll cut into the scrap paper and not your project. This means rotating your paper underneath your ruler as you work. This is a little more effort but 100 percent worth it to prevent accidentally slicing off bits of your project!

Making the Cityscape

Before you start cutting your cityscape, practice cutting different shapes using a combination of straight and curved lines. Remember to watch your hand position and respect your tools. With a sharp blade you don't need to press hard, just guide the blade over your material with light pressure.

Once you're comfortable with your blade, use your sketch as a guide, and carefully cut out your cityscape. Try not to use your sketch as an exact template, but see it more as a guide. Cut the outline of your cityscape first before cutting the details. When cutting your holes, make sure that you leave a decent margin of card for the "windows."

When you've finished cutting your cityscape, flip it over so that the back is facing up. Apply glue to the edges of the buildings, and stick your tracing paper over the top. From the front this will look like windows, diffusing the light from the LEDs. Trim off the excess tracing paper.

Finally, score a ¼-inch flap at the base of your design so that it folds forward (away from where the pop-up support will go).

How to Plan More Complex Paper Circuits

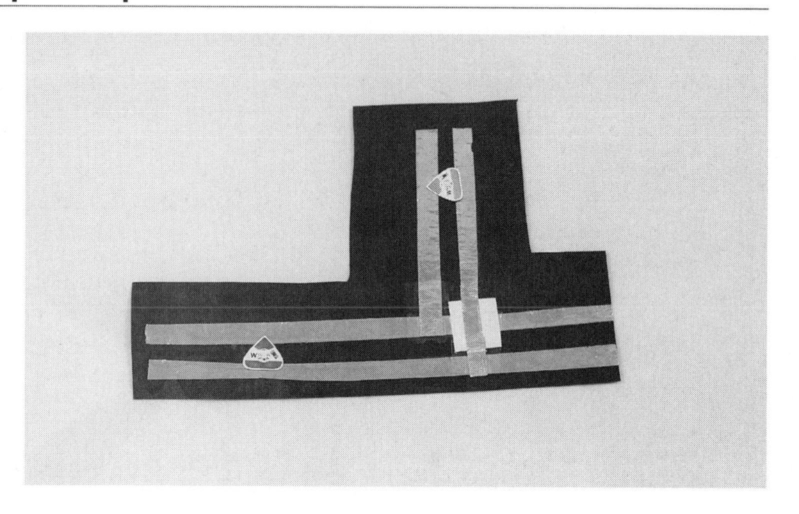

For this project we're using three Chibitronics LED stickers. Until now, we're used fairly simple circuits with straight lines, but you may want to take one of your stickers on a different path, for example, up one of your buildings. To do this, you'll need to know how to add an extra "branch" of a copper tape path and also how to build a "bridge" so that you don't cause a short circuit by touching positive and negative paths.

To make a branch from the main copper tape path, run your new bit of copper tape down to the main path to which you want it to connect. Rip or snip off the end of the copper tape with just over double the length you'd need to reach the main path. Then fold the tape in on itself so that the two sticky sides stick together. Do not snip off the end or you'll break the connection. Finally, use a tiny bit of clear sticky tape to stick this flap to your card on top of your original path.

Next, you'll want to connect the other side of your branch to your main copper tape path. You want to do exactly the same thing, but this time build a little bridge out of a bit of card. Card is not conductive, so it will insulate the two paths from each other, preventing a short circuit.

Starting Your Cityscape Circuit

Glue the bottom of your cityscape flap to your paper, right next to the pop-up cuts. The flap should hinge forward. Don't attach it to the pop-up cuts yet.

Now use copper tape to make your positive path. Make the main path first, and then bring your copper tape down to the paper. We're going to hide a DIY battery flap behind our cityscape, so make sure that you stick enough copper tape on the paper to fit a 3V battery before you snip it off. Leave about an inch of space on the paper to make sure that you've got enough room for the other path to come down and connect to the battery.

Next, make your first branch if you're using an LED in a skyscraper as in the picture. Start at the top, and run a new piece of copper tape down to the main path. Rip or snip off the end of the copper tape with double the length you'd need to reach the main path. Then fold the tape in on itself so that the two sticky sides stick together. Finally, use a tiny bit of clear sticky tape to stick this flap to your card on top of your original path.

Finishing Your Cityscape Circuit

Start making your main negative copper tape path. When you get to the branch of the positive path, you'll need to use a little bit of card to make a bridge. This will prevent a short circuit. Your sticky tape shouldn't cover too much of your path because you'll need to connect the negative side of your branch later.

Finish the main path first. Then bring your copper tape down to the paper, about ½ inch away from the positive path. Bring the path in line with where the battery will go. Then make a corner with the tape. Trim your tape to about 2 inches, and then pull the backing off and fold the remaining tape in on itself so that the two sticky sides stick together. This should leave you with a flap, which is where you're going to put your battery.

Next, make your negative branch. Start at the top, and run a new piece of copper tape down to the main path. Rip or snip off the end of the copper tape with double the length you'd need to reach the main path. Then fold the tape in on itself so that the two sticky sides stick together. Finally, use a tiny bit of clear sticky tape to stick this flap to your card on top of your original path.

Finally, add your LED stickers, making sure that the negative side of the stickers is connected to the negative side of the path and that the positive side of the stickers is connected to the positive side of the path.

Adding Power

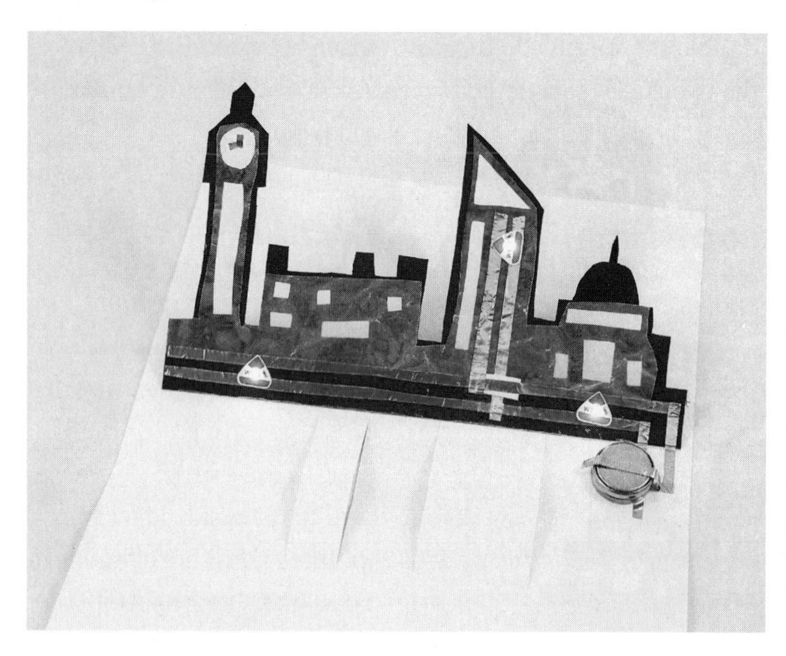

As we learned earlier, using more than one LED increases the amount of power being used in the circuit. We can also increase the amount of power available by linking batteries. We're going to look at how this works in a lot more detail in the next part, but for this project we're simply going to see what happens when we link two 3V batteries together in series.

To link two batteries in series, you need to link the negative end of one battery to the positive end of the other battery. Push one 3V battery into the flap with the positive side on the positive path. Touch the negative flap to the negative side of the battery. What happens? Now add the second battery into the flap, facing the same way as the first so that the negative side of the first battery and the positive side of the second battery are touching. What happens?

Be careful with this new knowledge: don't try adding more batteries without knowing a bit more about how it works or you could break your components! Finally, attach your pop-up cuts to the cityscape with glue or double-sided sticky tape, and admire your work.

Fix It

Not working? Don't worry! Follow these steps to figure out why, and fix it.

1. Check your power.
   - Is your battery the right way around? Flip it over and see what happens.
   - Has it run out of juice? Try another battery.

- Is your battery connecting into your circuit? Make sure that both the negative path and the positive path are connected to the correct side of the battery and that the positive and negative paths are not touching.

2. Check your components.
   - Are all your components the right way around?
   - Are all your components securely stuck in place? Loose connections mean that your circuit won't work.
   - Are any of your components on top of an unbroken bit of copper tape? Electricity likes to take the easiest route possible, so if you put a component on top of an unbroken bit of copper tape, the electricity will go through the copper tape instead of the component.
3. Check your wiring.
   - Do your positive and negative paths touch? If they are touching, your circuit won't work.
   - Are all your paths complete? If your copper tape path has rips or tears in it, your circuit won't work.
   - Have you made your path by sticking one bit of copper tape on top of another? Even if your glue is conductive, this can lead to a faulty circuit.
   - Do your branches connect properly? Make sure that they are stuck down to the main path securely. If you snipped off the end of the tape at the connecting end, your branch won't work.

Make More

For my next project, I'm making a cityscape of the city I just moved to, Berlin. Why don't you make a cityscape for all the different places you've lived or all the places you'd like to visit?

Maker Spotlight

Name: Coco Sato
Location: Brighton and London, UK
Home: Tokyo, Japan
Job: Artist and author

Coco is an award-winning artist, author, and performer whose practice reinvents origami in uniquely modern ways. Coco was born in Japan and educated in fine art at Central Saint Martins in London. She shares her art and skills via social media and video tutorials, performs at events, and creates creative campaigns for high-profile brands. Coco is the author of a book that combines the art of origami and electronics entitled *Creative Origami and Beyond*. As a child, she wanted to become an astronaut, to explore the farthest places she imagined. As an adult, she hopes to realize her childhood dream by collaborating with NASA and sending her art into space.

Q. ***What do you like to make and craft?***
A. I love using materials that are easily found around the home. I am particularly interested in fusing traditional and modern. I am fascinated with exploring history and heritage and how they have shaped the people of different cultures.

Q. ***What are you currently working on?***
A: I am collaborating with Ross Atkins, who invented The Crafty Robot, and experimenting with his new product, Smartibot. Smartibot is a simple robotics microcontroller based on an artificial intelligence (AI)–enabled iPhone app. Roborigami was originally developed by an engineer, Dr. Ad Spiers, and an interaction designer, Dr. Pete Bennett. However, with Smartibot, I can do so much without input from an expert.

Q. ***What do you want to do next?***
A. I love learning and sharing my journey with others. I am about to start lecturing at Ravensbourne University in London, which is very exciting! I'm also continually learning new skills and technologies. In the future, I would love to take a part in TED residency in New York City.

Q. ***What is your dream project?***
A. I would love to collaborate with NASA and send my art into space.

Q. ***Who inspires you?***
A. Artisans, scientists, engineers, and educators. I am inspired by people who are committed to making a positive impact on people's lives whether that is in the field of arts or science. I am inspired by fine details of everyday objects that are used for generations—items people continue to embrace because they are at once practical and beautiful. My work has taken me around the world. At every destination I investigate that culture's traditions, rituals, arts, textiles, and crafts.

Q. ***Can you share your favorite books, websites, or places you go to learn new skills?***
A. Definitely YouTube is the place where I go to find a lot of free tutorials. If you want to replicate some of my work, you can watch my video tutorials on my website, cocosato.co.uk.

Q. ***What is your favorite place to get materials or tools?***
A. With paper art, you don't need a special tool or skill to get started. Just grab copy paper or even an unwanted magazine. With a little imagination, lots of things you find around you can be used as material.

Website: http://cocosato.co.uk
Instagram: @giant_origami

Roborigami

Roborigami started in 2010 as a fun collaborative project between myself, robotics engineer Dr. Ad Spiers, and interaction designer Dr. Peter Bennett as an escape from our professional commitments. We combined traditional origami with the sound and movement of robotics to transform public spaces into playful Zen gardens.

Roborigami is a physical representation of my ongoing research—combining art, psychology, Zen philosophy, science, technology, and play to take you on a meditative journey inspiring connection and cultural exploration. It started with my sincere desire to inspire interest and foster understanding in different cultures. The goal is to dissolve boundaries and simply to make the world a fun place for all.

Gesture-Sensing Origami Fan

The gesture-sensing origami fan was born at a hack session with Helen Leigh, author of the book you're reading. At the time, we were both Maker in Residence at a makerspace in London called *Machines Room*, and we decided to spend the afternoon together with our favorite tools and materials to see what happened.

We got talking about nonverbal communication in different cultures, and I shared a story of how the different positions of a handheld fan can have different meanings in Japan. This was very inspiring and led to us creating a gesture-sensing origami fan for flirting without saying any words.

Using a tiny computer called a *micro:bit*, we made a fan and a bracelet that communicated with each other using radio signals. Depending on the movements of the fan, from being held in front of the face to fanning the face, the bracelet would receive different messages from "Come talk to me" to "Not interested, thanks!"

PART TWO

Soft Circuits

Introduction to Soft Circuits

Welcome to the Soft Circuits part. Soft circuits are a way of sewing with electricity. When most people think about electronics, they might think about wires and boxes, but by using your soft circuits skills, you can make the world of DIY electronics squishy, flexible, or wearable.

Soft circuits are electric circuits made using fabric. You might also hear them called *e-textiles* or *sewn circuits*. They are a mash-up of traditional textile techniques and electronics skills. You can sew all sorts of things into your soft circuit projects from simple LEDs to sensors, motors, buzzers, chips, or even tiny computers called *microcontrollers*.

This part will introduce you to the world of textile circuits. By the time you finish this part, you will have enough skills and knowledge to start designing your own simple soft circuit projects. If the idea of sewing with electricity is exciting to you, you'll be happy to know that the wearables part later in this book (Part Three) will take the soft circuit skills you learn in this part to the next level. By the time you finish this book, you will have all the skills you need to make exciting squishable electronics projects, design your own smart clothing, and create interactive accessories.

Over the course of this part, I'm going to guide you through the basics of sewing with electricity. I'll give you lots of tips and tricks for using conductive thread and sewable components in your projects. On the next page I'll explain some basic electrical sewing skills, and then later in this part you'll learn how to avoid basic mistakes, how to use common components, and how to make a DIY battery pack to power your projects. Each of these skills is an element of one of the five projects in this part, but if you want to skip ahead or look back, here's where to find this information on sewing with electricity:

Soft Circuits: Essential Skills

We will also take a closer look at circuits in this part. You will explore different types of circuits and find out what happens when you link up components or batteries in parallel and series. You'll also learn what switches and tilt sensors are and then make your own. Here's where to look up this information on how circuits work:

On the next page you'll find basic sewing techniques such as tying a knot, sewing a simple running stitch, using thread, and finishing your stitches. This part will also show you how to sew a backstitch and blanket stitch, two important craft techniques that will help you make beautiful and durable soft circuit projects. Here's where to look up information on these sewing skills:

Soft Circuits: Essential Skills

How to Thread a Needle

Cut a length of thread, and then moisten and squeeze one end so that it will fit through the eye of your needle. If the end of your thread is ragged, trim it with scissors. Hold the thread in one hand and the needle you want to use in the other. Slowly and carefully guide the thread into the eye of the needle. I find it easier to hold the needle sideways, meaning that I can't see the eye but I can see the thread as soon as it pushes through the eye. Pull the end of the thread through the eye of the needle until the end is 3 or 4 inches away from the eye of the needle. Tie a knot in the long end, and you're ready to start sewing.

To double thread your needle, pull the thread through until the two ends meet. Then tie a knot. Double threading means that your stitches will be stronger, but double threading when you're using conductive thread tends to tangle up more than single threading.

How to Tie a Knot

Loop the end of the thread around your forefinger once, and hold it in place with your thumb. Rub the thread with your thumb, rolling it toward the tip of your finger, and then carefully slip it off, keeping the loop intact. The thread loop should now be twisted around itself. Pull the ends of this loop into a knot. When using conductive thread, you can secure your knot in place with a dab of clear nail varnish.

How to Do a Running Stitch

The most basic stitch in sewing is called a *running stitch*. It's extremely useful and very easy to learn. Once you've threaded your needle and knotted the thread, push the needle through from the underside of your fabric. This means that the knot will be hidden, so your work will look neater.

Pull your needle and thread out and all the way through until it's stopped by the knot. Next, push the needle through the top side of the fabric back down to the underside and pull through until the thread is taut but not bunching the fabric. That's all there is to a running stitch: just repeat this process until your stitches have gone where you wanted them to go. For best results, keep your stitches small, neat, and close together.

How to End a Stitch

When you've finished sewing your projects, you need to tie your thread off. Insert the needle underneath a nearby stitch, and pull it under that stitch until it forms a loop. Push your needle through the loop and pull it tight. This will make a knot. To make your knot extra secure, repeat this process a couple of times. Once you're happy with your knot, trim the extra thread on the outside of the knot. Again, if you're using conductive thread, you should secure your knot in place with a dab of clear nail varnish.

Quick Start Tips

You'll find lots of good tips and tricks for sewing your soft circuits as you read through this chapter. Here are the most important things to remember:

- *Keep your stitches small, neat, and close together.*
- *Make sure that your components are sewed onto your fabric securely.*
- *Watch out for loose ends, fraying thread, rogue loops, or loose stitches.*
- *Conductive thread can be tricky to keep knotted. Secure your knots in place with a dab of clear nail varnish.*
- *Buy a special pair of scissors for working with fabric, and don't use them to cut anything else, even paper. Never use your fabric scissors to cut conductive thread or you'll blunt them!*
- *Practice new stitches and techniques on scrap material before using them in your projects.*

Tools and Materials List

To make all the projects in this part, you'll need the following tools:

- Two sharp pairs of scissors, one for fabric and one for everything else
- One pair of needle-nose pliers
- One 8-inch embroidery hoop
- Three 4-inch embroidery hoops
- One set of sewing needles
- One box of sewing pins
- One 12-inch ruler

You'll also need the following materials:

From a Craft Store

- Assorted colors of soft felt, including pink, yellow, and navy (You'll need about 10 squares of felt in total.)
- Assorted colors of ordinary cotton thread, including pink and yellow
- Assorted colors of embroidery thread, including white
- Buttons, beads, and pompoms
- Wadding

From an Electronics Store

- One reel of conductive thread
- Six sewable LEDs
- Nine 3-mm white LEDs
- One Teknikio star sewable LED
- One 3-mm yellow LED
- One sewable on/off switch
- One sewable buzzer
- Seven sewable 3V battery holders
- Ten 3V batteries

From a General Store

- Tracing paper
- Pencils
- Chalk
- Clear nail varnish

PROJECT 7

Circuit Sewing Sampler

Discover new techniques and start sewing with electricity

Tools

- Three embroidery hoops, at least 4 inches across
- One needle
- Sharp scissors
- Pencil and paper

Materials

- Three 6- × 6-inch squares of felt in any color
- Conductive thread
- Four sewable 3V battery holders and batteries
- Five sewable LEDs
- Clear nail varnish

Samplers are a way of displaying and recording needlework. They have been around for thousands of years, with the earliest remaining sampler, from Peru, dated at 200 BCE—that's well over a thousand years old! Until modern times, samplers were mainly used by professional needleworkers as a personal reference point. If you learned a new stitch or pattern, you would add it to your sampler so that you could remember how to do it later.

From the eighteenth century, samplers started becoming popular as an educational tool. If you were a wealthy young woman, you might display your intricate needlework in your family home to prove your skills and education to potential suitors. If you were a working-class young woman, your sampler might be long and thin, designed to be rolled up and taken to new employers to show how good you were at sewing, mending, and embellishing.

For this project, I decided to make three small samplers showing different circuit sewing skills. I chose fabric in different shades of the same color and framed them in their own embroidery hoops to hang on my wall. How are you going to design yours?

Choosing Your Materials

Start thinking about what you want your finished sampler to look like. What will you use it for? Do you want a long, thin, flexible sampler that you can roll up and take to your friends' houses to help them learn? Do you want to make and frame a neat square so that you can show people your skills when they visit your home? Do you want lots of small samplers in their own frames to decorate your bedroom? You could even make a really big sampler with plenty of room for adding new stitches, components, and techniques as you learn them in the future.

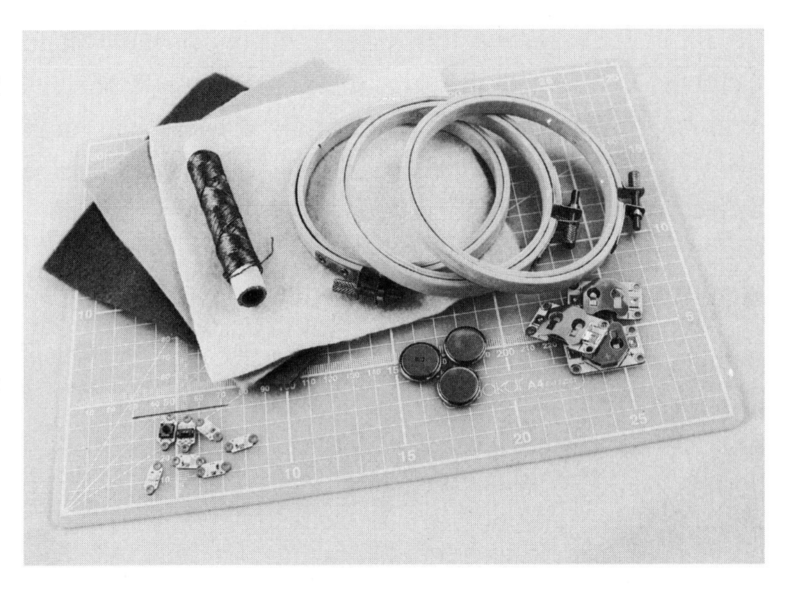

Once you've thought about the way you want your sampler to look, prepare your materials and think about your circuits. One of the cool things about sewing circuits is that you can easily take your thread on all sorts of different paths. Keep your first designs simple until you get used to working with conductive thread. Once you're confident, as long as you're careful about not crossing your positive and negative threads, you can let your imagination run wild!

Use your pencil and paper to start sketching out ideas. Play around with different versions. You can always change your mind when you start sewing, but I find it helpful to have a plan in my mind when I am working.

Preparing Your Sampler

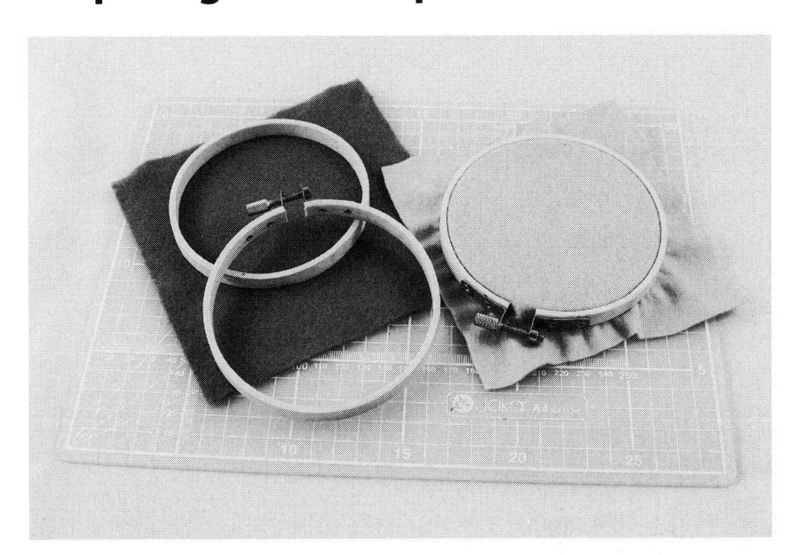

For this project, we're going to be using a tool called an *embroidery hoop*. Embroidery hoops hold your material firmly in place, making it easier to sew patterns onto fabric.

To put your fabric into your embroidery hoop, loosen the bolt, and slip the smaller ring out of the larger ring. Next, place your fabric over the top of the smaller ring, and then gently push the larger ring over the top of both. You may need to loosen the bolt a bit more to do this. Once the rings are back together with the fabric in between, you can gently tighten the bolt again to keep it all secure while you work.

Once you're finished sewing the area of your sampler inside your embroidery hoop, simply loosen the ring and move the hoop along to the next part you want to sew.

Thread one of your needles with conductive thread. Conductive thread can be slippery or fray easily, meaning that your knots could unravel. My tip is to dab a teeny tiny bit of clear nail varnish on the end of the knot. When the nail varnish dries, your knot will stay in place. Be careful, though: nail varnish isn't conductive, so make sure you're making a good connection with the nonvarnished part of the thread.

How to Make Good Connections

Before you get going on your sampler, I want to share some advice about sewing circuits. Being aware of the potential problems before you start will help you avoid them as you work.

There are two major things to watch out for. The first problem you'll need to be aware of is making sure that your connections are secure. Sewable components have handy holes that make it much easier to sew with electronics. When you're using these holes to sew your component onto your fabric, make sure that you make enough stitches to keep the component secure and the conductive thread firmly in contact with the component. This will also make sure that your component is securely fastened to your fabric.

Look at the pictures for two examples of sewn components. The example on the top is sewn nicely in place, but the stitches in the example on the bottom are too loose. This means that the electricity would not be able to get to the component, so this sewable LED wouldn't light up!

How to Avoid Short Circuits

Another problem you could have is with short circuits, which we introduced in the Paper Circuits part. As well as being careful not to cross our positive and negative paths, you need to watch out for loose ends.

Earlier, we put a dab of nail varnish on the end of our conductive thread knot. This is because some types of conductive thread are slippery. Some conductive thread also tends to fray. When conductive thread frays, even a tiny wayward strand could short circuit your whole project. To avoid this, try not to put the positive and negative paths too close to each

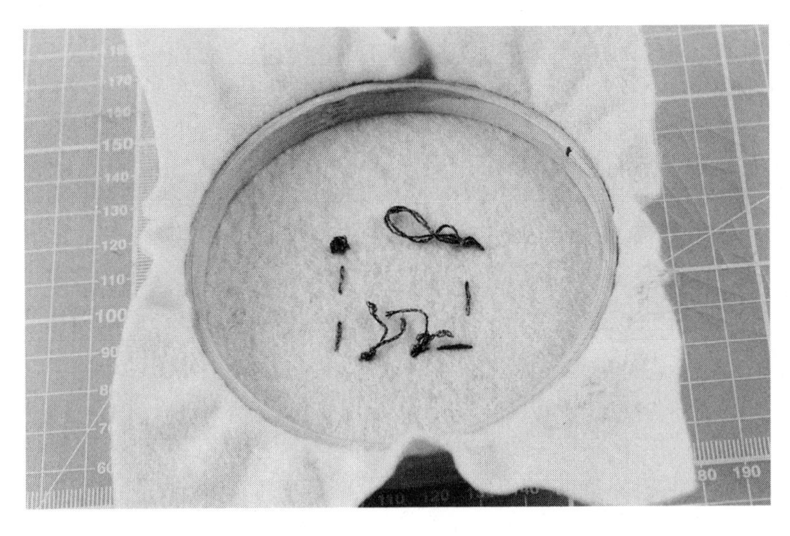

other. You can also use a light coating of hairspray to keep the strands in place once they are sewn.

The other way you might get a short circuit is by having loose ends of thread. While you're sewing, check the underside of your fabric for rogue loops, untidy knots, and loose stitches that could sabotage your circuit. The picture provides an example of loose threads that could stop a circuit from working.

My final tip before you start sewing is to try to keep your stitches small, neat, and close together. Not only will this look great, but it will keep your stitches in place when you move, bend, or stretch your fabric.

Sewing Your First Circuit

Place one of the sewable 3V battery holders on the fabric inside your embroidery hoop. Using your knotted and secured conductive thread, firmly stitch the bottom positive hole to the fabric and then the top positive hole. Either positive hole would make a connection to the battery, but we're sewing both to keep the component nice and secure on the fabric.

Next, we're going to sew the path. Because this is your first sewn circuit, we'll keep it simple and use just one LED. Before you start sewing, think about how you want your circuit to look. I've chosen a basic square shape so that you can see what path the electricity is taking, but you can choose to make your simple circuit a circle, a wavy path, or something more imaginative.

Once you've decided on your path, you can either mark your pattern with chalk or start sewing it freehand. Remember to try to keep your stitches small, neat, and close together. You should also regularly check the underside of your embroidery hoop to make sure that there are no rogue loops, untidy knots, or loose stitches.

Completing Your First Circuit

When you get to the part of your path where you'd like your LED to go, pause for a moment before sewing it on. Check that you are connecting the positive side of the sewable LED to the positive side of the battery, and make sure that you sew it on securely.

Once you've sewed the positive side firmly in place, tie your knot, secure it with a dab of clear nail varnish, and cut the thread neatly so that there are no loose ends dangling around.

Next, repeat the whole process again, but from the negative side of the battery to the negative side of your sewable LED. Make sure that your negative path stays well away from your positive path or your circuit won't work. I've made the negative side of my path a little bit wiggly, just for fun.

Once you've finished your negative path, your first sewn circuit is complete! Add a 3V battery into your battery pack. Does it light up? Give yourself a high five (some people call this clapping).

If your circuit doesn't light up, skip to the end of this project for a checklist of troubleshooting tips.

How to Make a Series Circuit

The next circuit we're going to add our sampler is called a *series circuit*. We have used series circuits before, in the Paper Circuits part, but here's a reminder of what they are.

When you wire (or sew) a project with more than one LED, there are two ways to do it. The first way is to wire them in series, and the second is to wire them in parallel. First of all, we're going to design and make a sampler using a series circuit. Then we'll make a parallel circuit in the next part of the project.

LEDs wired in series are connected end to end, meaning that the negative bit of the first LED connects to the positive bit of the second LED and the negative bit of the second LED connects to the positive bit of the third LED and so on. This whole daisy chain of LEDs should be connected up as usual to your power supply: negative to negative and positive to positive.

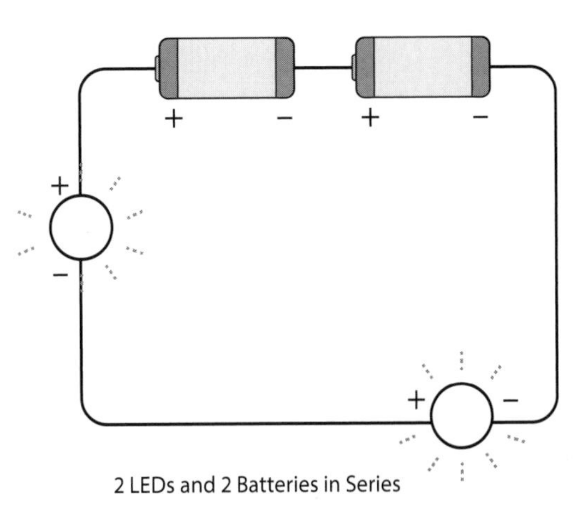

2 LEDs and 2 Batteries in Series

Adding in lots of LEDs or other components will change how much power you need to make your circuit work. For this circuit, we are going to wire up two LEDs and two battery packs, all daisy-chained together in series. We will learn more about how this works and how you can calculate the amount of power you need for bigger projects later in this book.

Designing Your Series Circuit Sampler

Get out your pencil and paper and start sketching some ideas for a series circuit sampler. You could keep it basic to show the daisy chain of a series circuit clearly, or you could add your own flair, taking inspiration from the shape of the circuit.

I've kept my series circuit sampler project fairly simple, but I also decided to add a crafty twist in honor of one of my favorite summertime insects: the lacewing.

As a child, I spent many happy summer vacations on my Aunt Mary's little farm in West Wales. As every farmer knows, there are some bugs that help you and some bugs that like to eat all your vegetables. Lacewings are not only a very pretty, delicate-looking insect, but they also eat up to 100 veggie-chomping aphids each day!

For my lacewing-inspired series circuit sampler, I chose a summery sky-blue background material. I designed the path of my conductive thread with wiggly stitches, like the flight path of an insect. The sewable LEDs represent the body of my lacewings, and I've chosen to add some lace "wings" to complete the sampler.

When you're designing your series circuit sampler, think about the same sorts of things. What color and type of material are you going to use? Guided by the daisy-chain shape of a series circuit, what path will your stitches take? Will you add anything? Remember to keep it simple while you're getting the hang of it.

Sewing Your Series Circuit Sampler

Once you've designed your series circuit sampler, you can either mark your pattern with chalk or start sewing it freehand using your sketch as a guide. You should also regularly check the underside of your embroidery hoop as you sew to make sure that there are no rogue loops, untidy knots, or loose stitches.

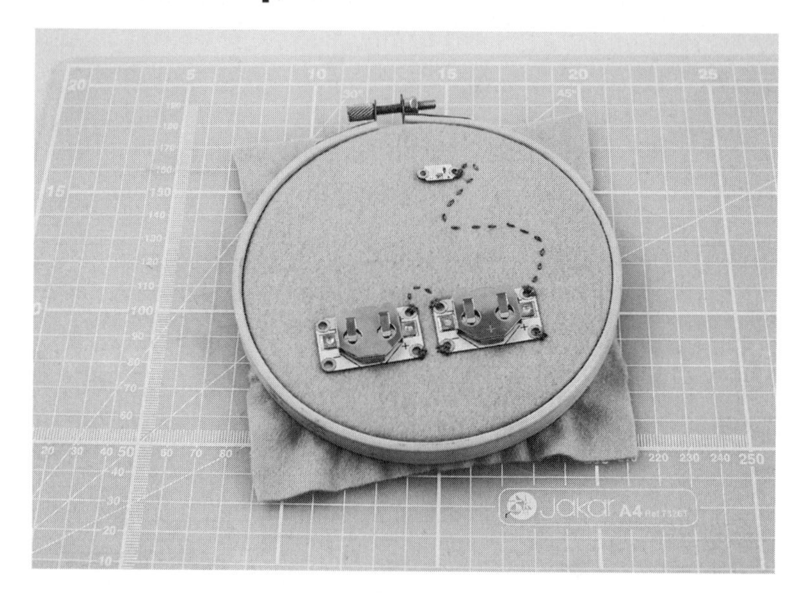

Double thread your needle with a length of conductive thread, and tie a knot. Secure your knot with a dab of nail varnish. Then start sewing your circuit from the positive side of the first sewable battery pack. Make sure that the holes are firmly stitched in place with your conductive thread. Then stitch your way to the negative side of the second battery pack.

Once the negative side of your second battery pack is sewn in place, tie off your thread and secure your knot. With a new piece of conductive thread, sew the positive side of your second battery pack to your fabric. Then start stitching your way to where you want your first LED to be. Before you sew it in place, make sure that you are connecting the positive side of the sewable LED to the positive side of the battery. Sew the LED securely in place to make sure that you have a good connection, and then tie your thread off and secure the knot.

Completing Your Series Circuit Sampler

Start again with a new piece of conductive thread at the negative side of your first LED. Thread your needle with a length of conductive thread, tie a knot, and secure it with nail varnish. As you start to sew the negative side of your first LED in place, take a moment to

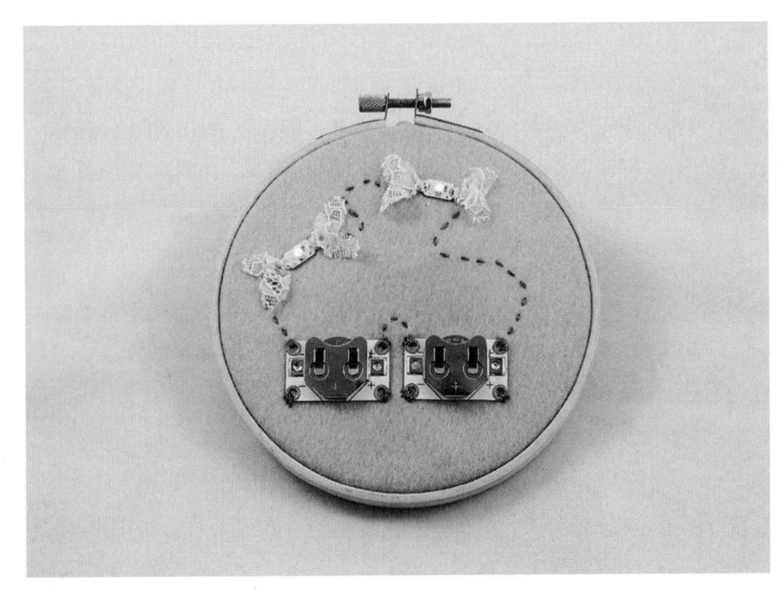

look at the underside of your series circuit sampler. Make sure that the knot isn't fraying or touching anything it shouldn't. Remember that your circuit won't work if two paths of conductive thread are connected.

Once you've sewn the negative side of your first LED to your sampler, take your second lot of stitches to where you want the second LED to go. Use your conductive thread to connect the positive side of the second LED with the negative side of the first LED. Tie your thread off, secure the knot, and start again at the negative side of your second LED. Take this third set of stitches back down to the negative side of your first battery pack to complete your circuit.

Put two 3V batteries into your battery pack for a moment to check that it all works. Then remove the batteries while you add the finishing touches that will bring your series circuit sampler to life.

If your circuit doesn't light up, make sure that you've sewn your daisy-chain series circuit in the correct way, and then go to the end of this project for a checklist of troubleshooting tips.

How to Make a Parallel Circuit

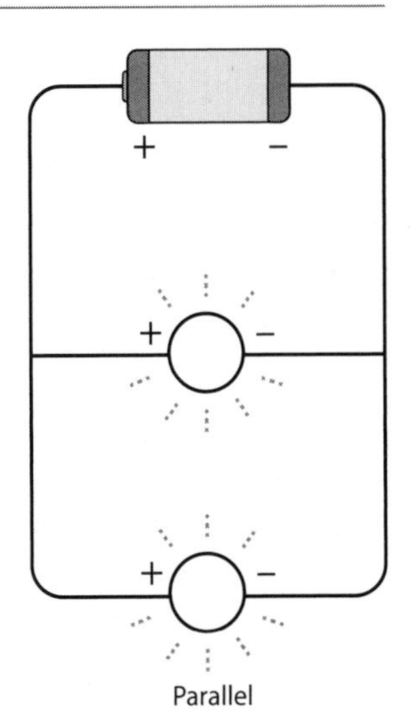

Parallel

Another way we can make a project with more than one LED is to wire (or sew) the LEDs in a parallel circuit. LEDs wired in parallel use one wire (or thread) to connect all the positive bits of the LEDs to the positive bit of your battery pack and another wire to connect all the negative bits of the LEDs to the negative bit of the power supply.

These two different ways of wiring circuits have different effects. Each component (e.g., an LED or a vibration motor) uses a certain amount of power, and each type of battery (e.g., 3V or AA) has a different amount of power that it is able to give. When you wire up two LEDs in series, the circuit divides the total power supply between the two LEDs. When you wire up two LEDs in parallel, each LED will receive the same amount of power.

In practical terms, a circuit with two LEDs and one battery wired in parallel will make your two LEDs shine just as brightly as the same circuit made with one LED, but it will also drain your batteries more quickly. A circuit with two LEDs and one battery wired in series may not work at all, or if it does, the LEDs will be much dimmer than if they were wired in parallel.

Designing Your Parallel Circuit Sampler

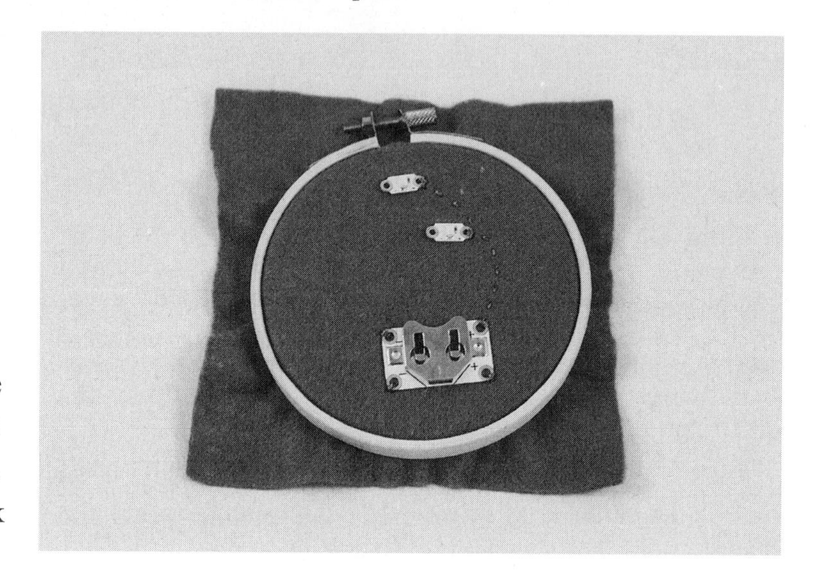

Get out your pencil and paper and start sketching some ideas based on the shape of a parallel circuit. You can also play with the layout of your components to find inspiration.

When I was making this sampler, I wanted an excuse to use some pretty blue and green sequins I just bought. These colors made me think of the sea, so I designed an abstract bubbly seascape. I chose a nice aquamarine square of felt that matched my sequins and contrasted nicely with the white LEDs.

I also knew that the aquamarine felt would look nice with the white and light blue colors of my other two samplers.

When you're designing your parallel circuit sampler, think about the path that your stitches will need to take. Keep it simple, and use only two LEDs. Remember that the positive and negative paths of the circuit cannot touch or your design won't work at all!

Making Your Parallel Circuit Sampler

Once you've designed your parallel circuit sampler, you can either use chalk to mark your pattern or just start sewing. As usual, thread your needle and secure the knot with a dab of nail varnish. As you sew, remember to check for rogue loops, frayed knots, or loose stitches.

Start sewing your parallel circuit from the negative side of the sewable battery pack. Then stitch your path to the first LED in your design. When you stitch your LED in place, make sure that you connect the negative side of the sewable LED to the negative side of the battery. When your first LED's negative bit is sewn securely in place, do not tie off your

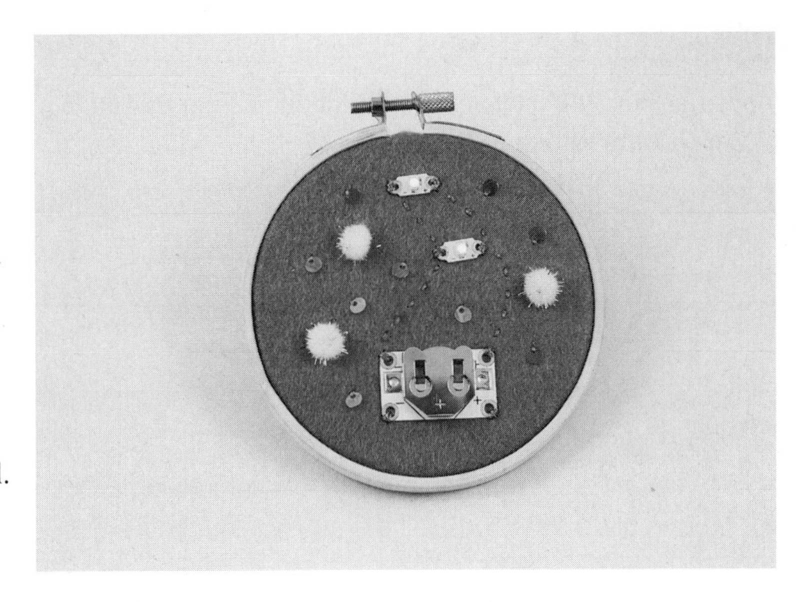

thread. Instead, carry on stitching until you reach the second LED. Sew the negative bit of the second LED in place, and then repeat the process for your third LED if you have one. Once you've finished sewing all the negative bits of your LEDs, tie off your thread and secure the knot as usual.

Next, start the process again with a new piece of conductive thread on the positive side of your battery pack. This time stitch all the positive bits of the LEDs together. Once you've tied off, secured your knots, and checked for any stray thread or loose connections, add your 3V battery and admire your work.

Fix It

Not working? Don't worry! Follow these steps to figure out why, and fix it.

1. Check your power.
   - Is your battery the right way around? Flip it over and see what happens.
   - Has your battery run out of juice? Try another battery.
   - Is your battery connecting into your circuit? Make sure that your battery fits snugly into the holder and that the conductive thread is sewn securely to the correct side of the battery.
2. Check your components.
   - Are your LEDs working? It's a good habit to check each component before you add it into a circuit.
   - Are your LEDs sewn securely in place? Loose connections mean that your circuit won't work. Tighten your connections and try again.
   - Did you sew each LED the right way around? If an LED is facing the wrong way, it won't work. Check carefully to see if your LEDs are facing in the right direction.
3. Check your wiring.
   - Do you have a short circuit? If your positive and negative paths touch, no matter how slightly, your circuit won't work. Tidy up your loose ends, check your knots for fraying, and restitch any crossing paths.

Make More

As you learn more and more techniques over the course of this part and beyond, think about adding them to your sampler. You can also fill your sampler with experiments—new stitches, new components, and new fabric samples. Your sampler doesn't have to include just circuit patterns. You can add new embroidery stitches or experiments with different thread and embellishment techniques with sequins or beads.

PROJECT 8

Squishable Sparkle Heart

Use your new conductive sewing skills to make an electric emoji

Tools

- Sharp scissors
- Needle-nose pliers
- Three needles
- Sewing pins
- Tracing paper and a pencil

Materials

- Two 5- × 5-inch squares of soft felt in pink and yellow
- One 3-mm yellow LED
- Pink and yellow cotton thread
- Conductive thread
- One 3V battery
- Wadding

Who doesn't love an emoji? The first one was invented by Shigetaka Kurita in Japan in 1999, and now we've got nearly 3,000 official emojis from which to choose. Across the world, over 60 billion emojis are sent on every day—and that's just on Facebook!

My favorite emoji is the sparkle heart, so I thought it would be fun to make one in real life. In this project, I'll show you how to cut out a simple pattern and make ordinary LEDs into sewable ones. I'll also share my top tip for making sure that you're sewing the positive and negative threads correctly as you sew so that you don't have to wait until the whole project is finished to check your connections.

Preparing Your Materials

Make sure that you've got all your tools and materials ready. Then prepare your felt. See page 249 for a traceable template for all the parts you'll need to make the sparkling heart emojis.

You can mark a pattern onto fabric in a number of

different ways. If I'm designing my own projects, I often sketch my ideas on paper and then trace my patterns onto the fabric using pins, tailor's chalk, or water-soluble fabric pens. I'll show you how I do this later in this part, but for this project, we'll be keeping it super simple with a traceable paper pattern.

Take some paper and place it over the template on page 249. Using a pencil, draw or trace all the different parts, and then use a pair of scissors to carefully cut them out. Next, place the paper parts onto the felt (pink for the heart and yellow for the stars and battery pack), and secure them in place using a sewing pin. You can then carefully cut around the shapes before removing the paper. The stars can be fiddly, so my advice is to cut a tight square around the star first, and then snip away five little triangles to leave you with a neat star. Don't worry if it takes you a few tries to get your perfect hearts and stars!

Sewing Your Hearts Together

Thread a needle with pink thread, and tie it off. Take one of your felt hearts, and pull the thread through the point of the heart. Place the second heart on top of the first so that it hides the knot. Then sew them together along one edge using a running stitch. We'll be stuffing our heart with wadding to make it squishable, so you'll need to keep your stitches small, neat, and close together. Stop sewing before you go around the first curve at the top of the heart. Leave your thread attached, and tuck your needle into the felt for later.

Now we're going to attach the stars. Double thread a needle with yellow thread, and tie it off. Flip your heart around until the pink stitching in on the right-hand side. Put your big star in the top-right corner, and then sew it in place by making a few stitches around the body of the star. Remember to make your first and last stitches from the inside to hide the knots. Do the same for the little star on the other side of the heart.

Next, we're going to make a couple of holes to poke the legs of the LEDs through the felt. I've found that the neatest way to do this it to use a needle to poke a hole through the star and heart and then wiggle it around a bit to make a little hole. You can then (gently!) push the legs of the LED apart a little bit and push one though the hole you just made. This can be a little fiddly, but take your time, and don't force it through or you might break the LED. Do the same for the other leg of the LED.

How to Prepare an LED for Sewing

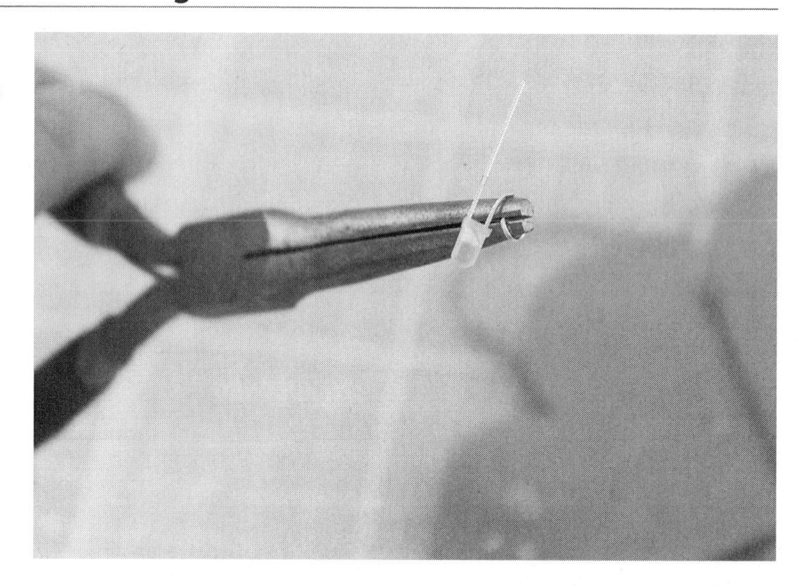

You should now have your stars and LED in place. The next step is to prepare your LED for sewing. Wrap each of the LED legs around the end of your needle-nose pliers into a circle. When making your circuit, you'll need to know which leg is positive (the long leg) and which leg is negative (the short leg), but it can be tricky to tell when they're wound up.

If you look closely, you can see that one edge of the LED bulb is flattened: this is the negative side. I still find it tricky to tell using this method, especially when I'm pushing LEDs through fabric, so my tip is to put more of the positive leg in the "teeth" of the pliers before you start turning it. This means that your positive leg will have a circle with a cross through it, and the negative leg will just have a circle.

This technique for making ordinary components into sewable components will come in handy for all sorts of sewable projects in the future, whichever method of telling the legs apart you prefer.

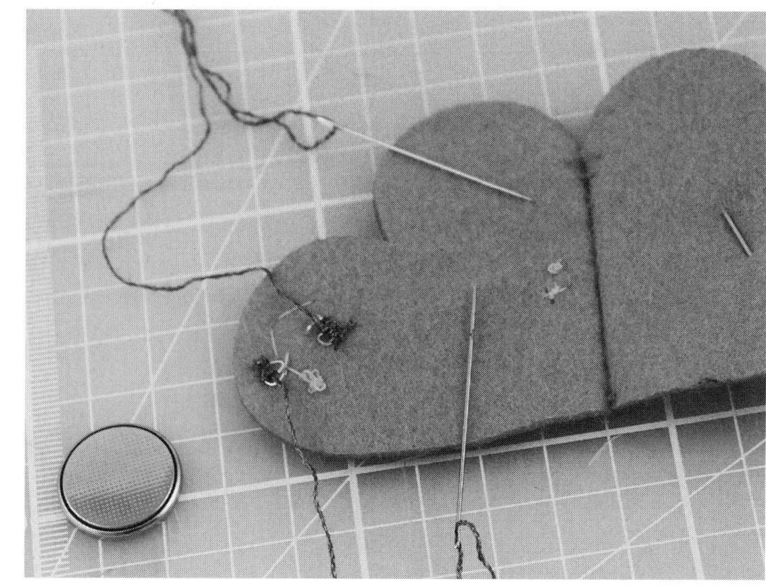

Anchoring Your LEDs

Thread a needle with a length of conductive thread, and tie it off. Remember to secure your knot so that it doesn't unravel in your finished project. The first thing to do is to anchor your LED. Starting from the inside of the hearts to hide the knot, stitch your LED to the felt, and then make four or five tight loops around the circular LED legs. These loops should be nice and secure: they're making the connection, so be sure that you've done this neatly. Once you've secured one leg, leave the needle and thread attached, and move on to the other leg.

Thread a second needle with a length of conductive thread, and tie it off. Repeat the anchoring with the other LED leg, making sure to avoid touching the positive thread to the negative thread.

At this point you should check that you've anchored your LED correctly and that everything is working. The reason I use two needles is that it makes it easy to check my circuit at any stage in my sewing: simply touch the needle connected to the positive leg to the positive side of a 3V battery at the same time as you touch the needle connected to the negative leg to the negative side of the battery. This technique has saved me so much time: if I get something wrong, I know early on instead of waiting until the end of a project.

Sewing Your Circuit

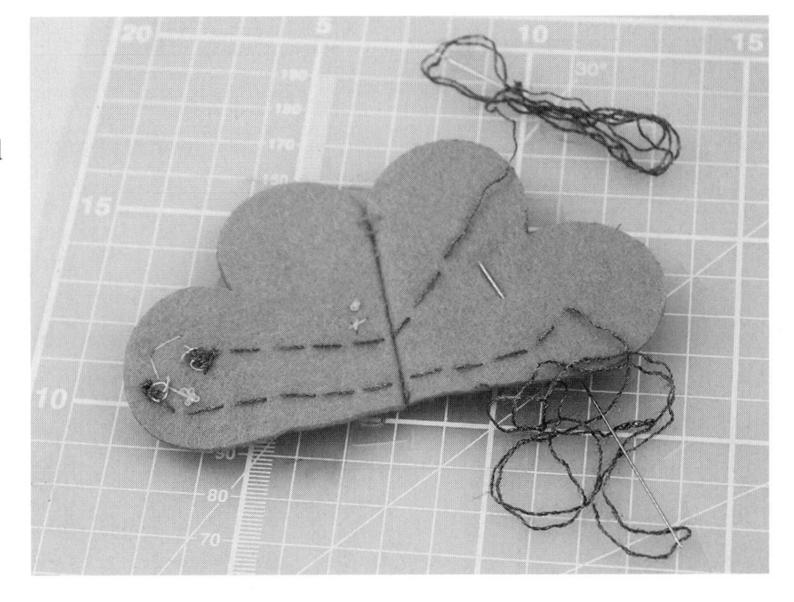

Now that your LED is anchored, it is time to sew the circuit. Use a running stitch to sew the positive and negative conductive threads along a path to the back of the heart, just as in the picture. You can take any path you like, but the paths must not touch. My positive path is above my negative path, but it doesn't matter which way your paths go. You just need to make sure that you line the battery the right way up later.

If you don't like the way conductive thread looks on the outside of the heart, you can make very small stitches on the outside and bigger stitches on the inside. If you have a very steady hand, you can also sew the circuit without pushing the needle all the way through the felt material.

Remember that you'll be stuffing your heart at the end to make it squishable, so cross over from the front heart to the back heart near the edges to stop the thread from getting in the way.

When you've finished sewing your circuit, the positive and negative threads should end up about 1 inch apart in the middle of the heart. Check that your circuit still works by touching your needles to a 3V battery.

Adding Power to Your Circuit

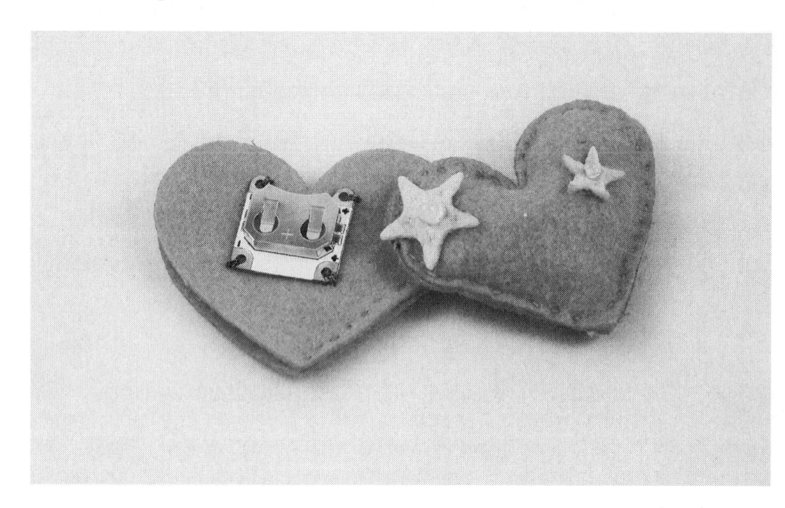

At this point, you have two choices. If you like, you can simply use your conductive thread to sew a small battery holder to the back of the heart, completing the circuit. If you choose this option, all you have to do is check that you are sewing the negative side of the LED to the negative side of the battery pack and vice versa. When you finish, make sure that you secure your knots.

You can also choose to try something a little more advanced by making your own battery holder. If you want to do this, cut out two of the tabbed circle shapes on page 249 in either pink or yellow felt. Then move ahead to the next step to find out how to put them together.

How to Make a Fabric Battery Pack

Making a fabric battery pack is a little trickier than using a ready made one, so be prepared to try it a couple of times before you get it right. However, once you get the hang of it, you'll never need to buy a 3V battery pouch for your fabric projects again! The main things to watch out for are that your positive and negative threads don't touch each other and that your pack is snug enough to make a good connection to the battery.

To start, push one of your conductive thread needles through to the outside of the heart. Push it through a tab on one of the fabric battery packs you cut out earlier. Secure the tab in place with a couple of stitches.

Next, use a running stitch to sew a path to the center of the tabbed circle, and then make several stitches in a rough star shape, just as in the picture. This star shape will be where one side of your battery connects.

Push your needle through to the inside of your heart, and then tie your thread off tidily. Secure the knot with a dab of nail varnish to prevent it unraveling and creating a short circuit.

Next, place the second tabbed circle shape on top of the first. Push your other conductive thread needle through both layers of tabs on the opposite side to the first tab you sewed. Secure the tab in place with a few stitches, and then sew a path to the center of the tabbed circle. Make sure that you only sew on the top layer, avoiding the bottom tabbed circle that you've already sewed. When you reach the center of the tabbed circle, make several stitches in a rough star shape. Tie off your thread, and secure the knot with a dab of nail varnish.

When you close this fabric battery pack, the two stars will connect with the positive and negative sides of the battery, allowing the electricity to flow through the circuit.

To finish your fabric battery pack, you need to thread a needle with ordinary cotton thread. Place a 3V battery between the two layers of tabbed circles so that you know the size, and then use your cotton thread to sew one tab first and then in a semicircle around the bottom of the pouch to the other tab. Sew the second tab in place, and test the size of the battery holder with your 3V battery. If the pouch is too loose, sew a little up the sides until they are snug but still removable. When you're happy with your DIY battery pouch, tie off your thread.

Finishing Your Squishy Heart

Now that you've finished the most difficult part of the project, it's time to sew up the rest of the heart. If you have chosen to make the fabric battery pack, you'll need to finish sewing that first. Put a 3V battery between the two layers of your battery pack, and sew a U shape around it using ordinary cotton thread. The battery should be nice and snug but not so tightly sewn in that you can't replace it later. Check that it all still works, and then remove the battery and start sewing the rest.

Once your battery pack is in place, go back to your first needle with the pink thread. Continue sewing around the edge of the heart until you have only about 1 inch left to go. Take a small amount of wadding and gently push it through the gap you've left in your heart. Keep adding wadding until your heart is squishy enough. Then sew up the gap, and knot your thread.

That's it! Reinsert the battery, give your extra sparkly sparkle heart a squish, and feel proud of yourself.

Fix It

Not working? Don't worry! Follow these steps to figure out why, and fix it.

1. Check your power.
   - Is your battery the right way around? Flip it over and see what happens.
   - Has it run out of juice? Try another battery.
   - Is your battery connecting into your circuit? Make sure that your battery fits snugly into the holder and that the conductive thread is connecting to the correct side of the battery. If you made your battery pack, make sure that your stitches are neat and not touching the other path.
2. Check your components.
   - Is your LED working? It's a good habit to check each component before you add it into a circuit.

- Is your LED securely sewn in place? Loose connections mean that your circuit won't work. Tighten your connections, and try again.

3. Check your wiring.

- Do you have a short circuit? If your positive and negative paths touch, no matter how slightly, your circuit won't work. Tidy up your loose ends, check your knots for fraying, and restitch any crossing paths.

Make More

If you want to go big, you can scale up this squishable heart project so that it's more like a cushion than a little plushie. If you want to make more projects inspired by this one, try designing and making your own favorite emoji. When I've made emojis with other people, their favorite emojis have been the robot face, the unicorn face, and the poop.

When you're designing your own emoji, keep your circuit simple and include only one LED until you've learned about sewing parallel and series circuits. Keep your emoji designs simple, too. Try not to use more than two or three layers of felt, and experiment with your patterns by cutting out paper trials first.

PROJECT 9

Tiny Squishy Torch

Light up your world with DIY pressure sensors

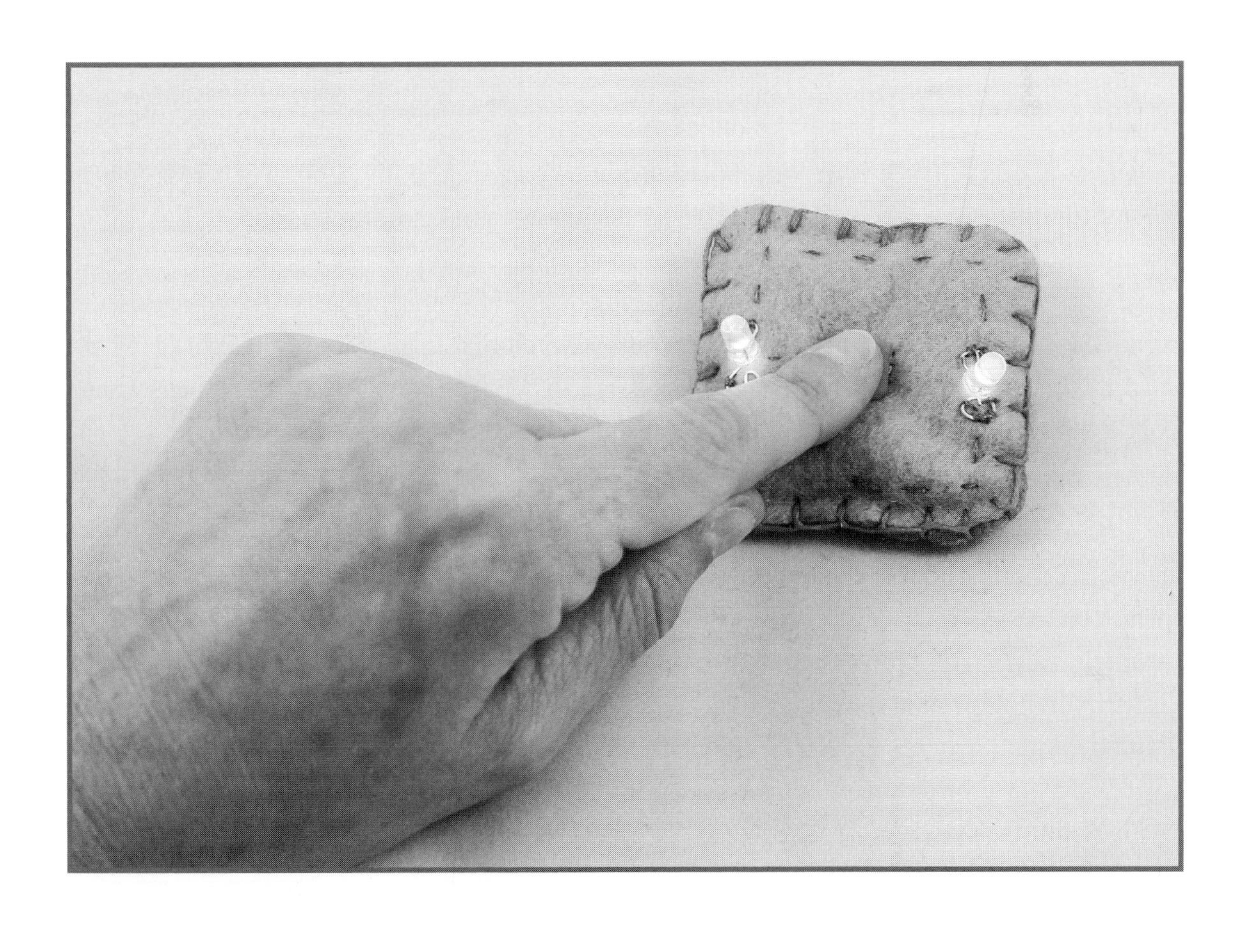

Tools

- Sharp scissors
- Needle-nose pliers
- Two needles
- Sewing pins
- Tracing paper and a pencil

Materials

- Two 5- × 5-inch squares of felt
- Two 3-mm white LEDs
- Cotton thread
- Conductive thread
- Two 3V batteries

Lights out everyone! Or is it? We all like staying up past our bedtime, but if you're trying not to wake everyone else in the house up, it's handy to have your own torch to light your way to the fridge.

We've come across switches a few times in this book, and in this project we're going to explore how they work and make our own squishy switch that turns your torch light on when you apply pressure and off when you release. We're also going to explore what happens when we link batteries in series and learn a new embroidery technique: the *blanket stitch*.

What Is a Switch?

A switch controls whether a circuit is open or closed. As you learned in the Paper Circuit part, an open circuit allows electricity to flow around it, but a closed circuit does not. Switches give us a handy way to turn our projects on and off without having to put the battery in or take it out every time we want to turn it on or off.

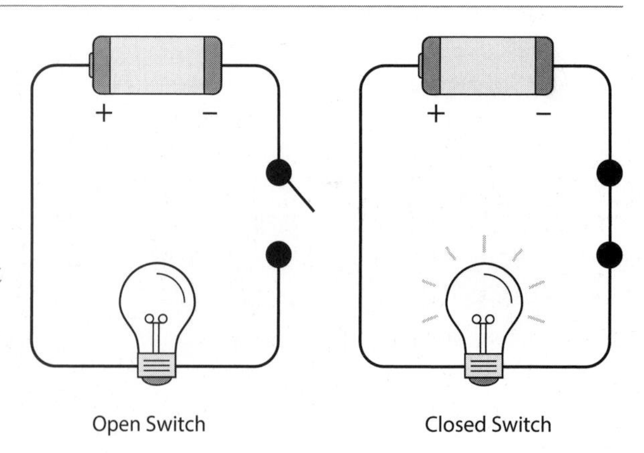

We used our first switches and buttons in the Paper Circuits part. There are many different types of switches that you can use in your projects, but they all fall under two broad categories: *maintained switches* that stay the way you left them in (e.g., a light that you switch on or off) and *momentary switches*

that only work when you're pressing on them (e.g., keys on a keyboard or the button of a doorbell).

In this project, we're going to make a momentary switch that responds to pressure, and then we are going to use that switch to make a tiny torch that lights up when you squeeze it.

Preparing Your Materials

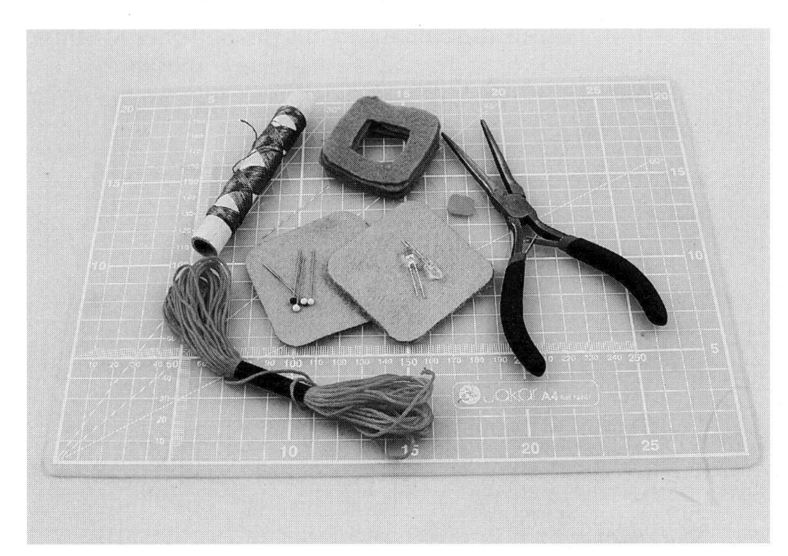

Make sure that you've got all your tools and materials ready. Then prepare your felt. Turn to pages 251–252 for a template of the tiny torch pattern. Trace and cut out the two bigger squares with rounded edges first. Pin them to one of your squares of felt, and cut the shapes out. These big squares are the outside of the torch.

You'll need to trace and cut out five of the smaller squares. You'll see that they have an even smaller square in the center. Cut that out, and discard the center, leaving you with a frame. These frames will be the center of your pressure switch. The remaining two shapes will be your battery pack and an optional small circle to cover your pressure pad stitches on the front of your tiny torch.

Thread a needle with conductive thread, and tie it off, securing the knot with a dab of nail varnish.

Sewing Your Pressure Sensor

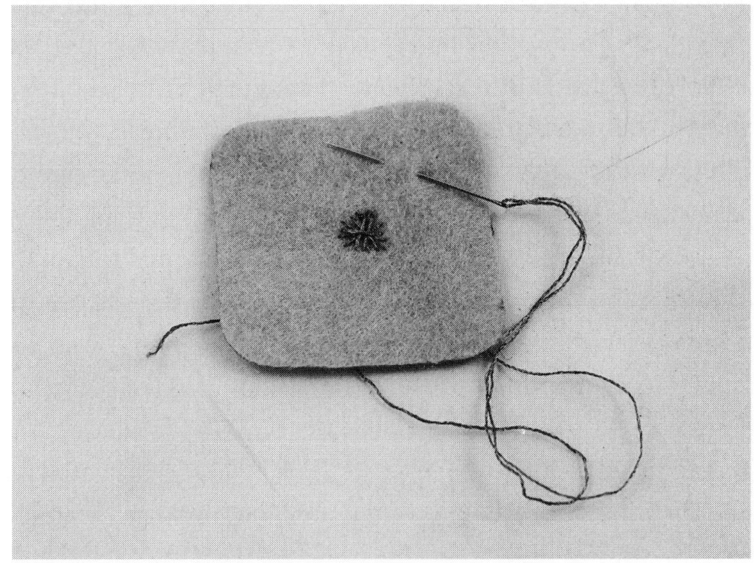

Take one of the larger squares of felt, and push your needle through the center. Make lots of stitches in a star shape covering a ½-inch area. This area of conductive thread will be the connection for your pressure sensor switch, so make sure that the coverage is good. Take a look at the picture to see the kind of coverage you should be aiming for. I made about 15 stitches in this example.

When you've finished your stitches in the center, push the needle through to the other side, and flip your felt over. Your stitches might look messy from this side, but that's okay: we'll cover them up later. This side is now the front of your project.

Now we're going to sew the LEDs of your torch. We're going to do this in a series circuit, like a daisy chain. When you wire up a component in series, you have to remember to add up their power requirements. We will talk more about this when we add the batteries for this project.

Sewing Your LEDs

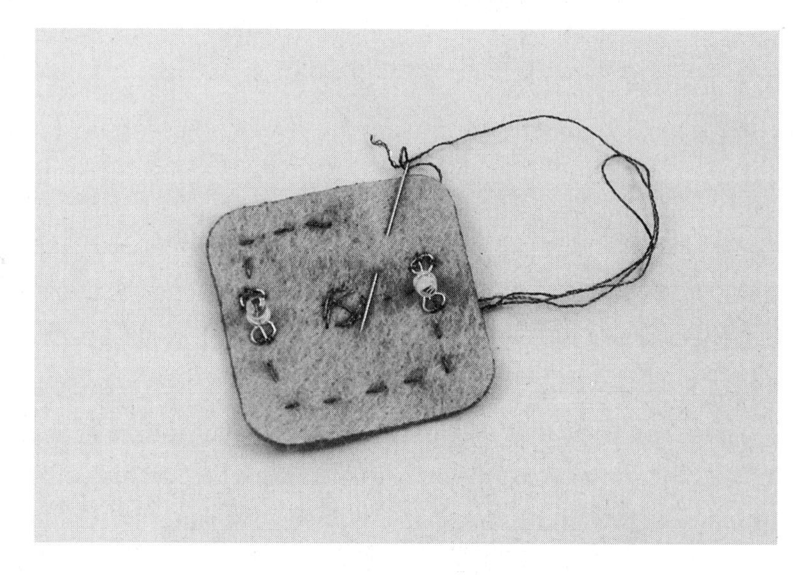

From the center star, stitch outward at a right angle to one of the sides of the square until you're about ½ inch away from the edge. Now we're going to add our first LED. For this project, I decided that it would look cooler to have the curly legs of the LED on display. Use your needle-nose pliers to make your ordinary LEDs into sewable LEDs. Remember to put more of the positive leg in the "teeth" of the pliers before you start turning it so that you can tell which leg is positive (the long leg) and which leg is negative (the short leg).

Once your LED is ready to sew, stitch the positive leg onto the felt. Tie off on the underside of the fabric, and secure the knot. Rethread your needle with conductive thread, and stitch the negative side of the LED to the felt. Take care that you do not overlap any positive and negative stitches or your LED will not light up. Once the negative leg is in place, sew down the side of the square, keeping about ½ inch from the edge. When you get to ½ inch from the bottom of the square, turn the corner and sew all the way along until you're ½ inch away from the next edge. Turn that corner with your stitches, and start sewing back up the side of the square. Just before you hit halfway, pause and prepare your second LED for sewing. Sew the positive side of the LED to the felt, and then tie off and secure the knot on the underside of your fabric. Rethread your needle with conductive thread, and stitch the positive side of the LED to the felt, making sure that you do not overlap any positive and negative stitches.

Sewing the Other Side of the Pressure Sensor

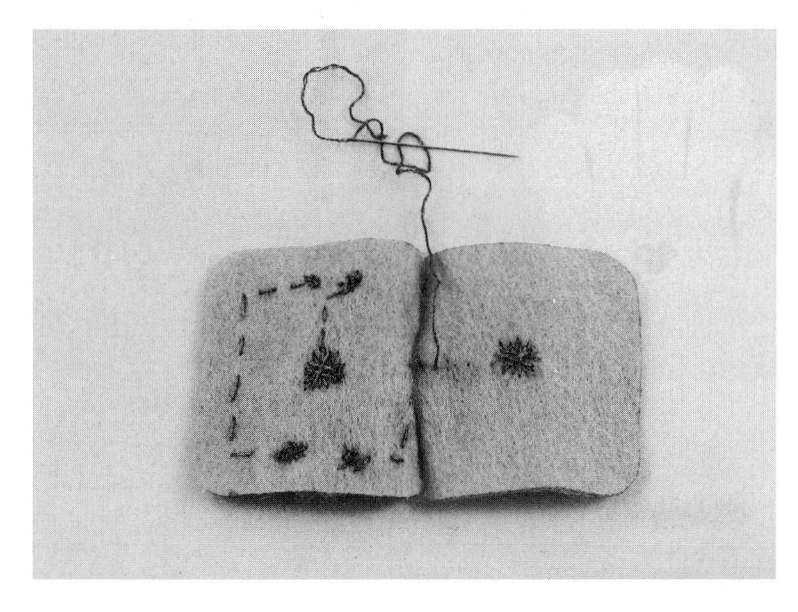

Sew your way from the negative side of the LED to ½ inch from the top of the felt square. Then pause your sewing, leaving your needle on the top of the fabric, to prepare the other side of the torch. Take a sewing pin, and pin the top of the two squares together, just under ½ inch away from the edge. Open the pinned squares out so that they form a rectangle, and start sewing again. Push your needle from the front square to the back square. Don't pull this thread too tightly because you'll need a little wiggle room when you add the pressure switch padding.

On the back part of your tiny torch, sew one or two stitches along the top edge of your square about ½ inch from the edge. You will be sewing pretty close to your pin, but take care to sew only through a single layer of felt: these stitches should be on just the back square. When you've reached the center of the top side of the square, pause your sewing of this path.

Thread another needle with conductive thread, and push it through the felt in the center of the square so that the secured knot is on the outside. Flip your squares over, and make another set of stitches in a star shape covering a ½-inch area. The two squares should look like a reflection of each other, as in the picture. Take your thread back to the other side of the material, tie it off, and secure the knot.

Sewing the Rest of the Battery Pack

Take the needle attached to the top of the square, and push it through one of the tabs on the battery pack bit of material. Sew a path to the center of the tabbed circle without going back through the square. Then make several stitches in a rough star shape. Tie off your conductive thread, and secure your knot.

Next, thread a needle with ordinary cotton thread, and secure the battery pack in place onto the back square. Sew one tab first and then in a semicircle around the bottom to the other tab. Sew the second tab in place, and test the size of the battery pack with two 3V batteries. If the pouch is too loose, sew a little up the sides until they are snug but the batteries are still removable. When you're happy with your DIY battery pouch, tie off your thread.

Finishing Your Pressure Switch

Stack your five layers of the small frame-like squares so that all the holes are aligned. Flip your squares so that the inside stars are showing, and place your frames between the two squares.

Take a moment to trace the path from the center star on the top, around the LEDs, over to the other side, through the batteries, and out to the star on the inside of the back square. The frame-like squares in the middle of your torch are keeping the two stars from touching, meaning that your circuit is not complete. When you squeeze the torch in the center, the stars touch, completing the circuit and allowing electricity to flow through and light up your LEDs.

If you like, you can now add the remaining circle from the pattern to cover the stitches on the front. Just make sure that your cotton knots don't prevent your conductive stars from making contact.

The next step is to sew up your project. You may wish to add your batteries at this point to test whether your circuit works. To see how to do this, skip ahead to the "Adding Your Batteries in Series" section. Make sure that you take your batteries out again before you sew everything up. When you sew the edges up, you can use a running stitch with small, tight stitches to keep everything in place. You can also use a technique called a *blanket stitch*.

How to Sew a Blanket Stitch

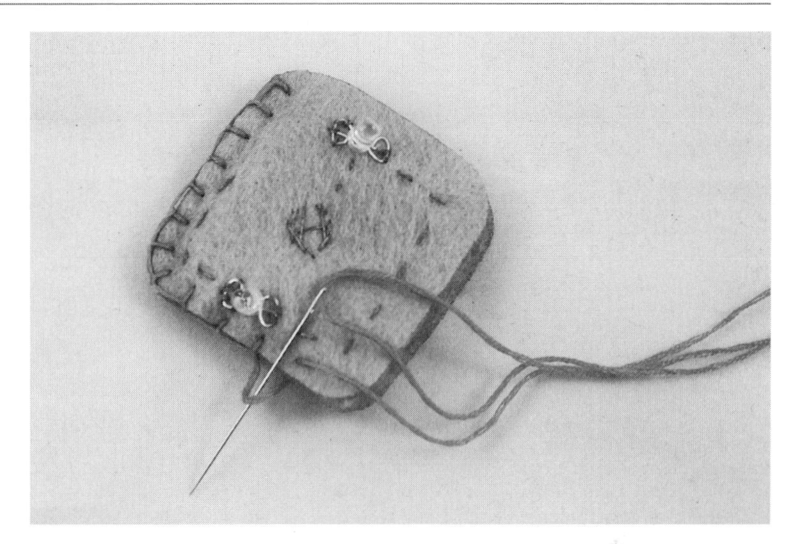

Pin all three edges of your torch to make it easier to sew, and double thread your needle with a long length of thread. I used two threads of embroidery thread about 40 inches long (20 inches long and four threads after doubling) in a contrasting color to my material. If blanket stitching is new to you, take some time to practice it on scrap material before starting to sew your torch.

Start your blanket stitch on the right-hand side of the unpinned edge, from the inside so that the first knot is hidden. Push your needle up to the top of your fabric about ¼ inch from the edge. Pull the needle around to the back of the torch, and push it through to the same place you started, making a loop around the edge. Before you pull the loop closed, send your needle from right to left under the loop stitch you just made, and then pull the thread through so that it is secure but not too tight.

Next, poke your needle from back to front about ¼ inch to the left of your first stitch. As you pull that thread through, you'll see another loop. Again, push your needle through the loop from right to left under the stitch you just made, and then pull the thread through. This is blanket stitch! Keep going counterclockwise around your torch, pulling out the pins when you get to them. To get the best results, keep the space between each stitch the same (about ¼ inch), and do the same for each stitch distance away from the edge (about ¼ inch). If you're finding blanket stitching tricky, search online for a video to help you with this technique.

Adding Your Batteries in Series

As we learned earlier, we can increase the amount of power in our circuit by linking batteries. In Project 10, we will link our batteries in parallel. For this project, we're going to bump up our voltage by linking two 3V batteries together in series.

If you wire two batteries in series, you increase the voltage available. For example, linking two 3V batteries in series will give you 6V. This is too much for one LED, but we've sewn in two LEDs in series. This means that the power required by the LEDs is added together, so 6V is approximately right. To link two batteries in series, you need to link the negative end of one battery to the positive end of the other. In Project 7, we did this by daisy chaining two battery packs together. In this project, we're simply going to stack two 3V batteries in the DIY battery pouch.

Push one 3V battery into your pouch with the positive side facing the front of the torch, and give it a squeeze. What happens? Now add the second battery into the pouch, facing the same way as the first so that the negative side of the first battery and the positive side of the second battery are touching, just as in the picture. Now your circuit should light up when you're squeezing it.

That's it! Happy squishing.

Fix It

Not working? Don't worry! Follow these steps to figure out why, and fix it.

1. Check your power.
   - Are your batteries the right way around? Flip them over and see what happens.
   - Have your batteries run out of juice? Try another set of batteries.
   - Are your batteries connecting into your circuit? Make sure that your batteries fit snugly into the pouch and that the conductive thread is connecting to the correct side of the batteries. Check that your stitches are neat and not touching the wrong path.

2. Check your components.
   - Are your LEDs working? It's a good habit to check each component before you add it into a circuit.
   - Are your LEDs securely sewn in place? Loose connections mean that your circuit won't work. Tighten your connections, and try again.
3. Check your wiring.
   - Do you have a short circuit? If your positive and negative paths touch, no matter how slightly, your circuit won't work. Tidy up your loose ends, check your knots and thread for fraying, and restitch any crossing paths.

Make More

DIY pressure switches are a great technique to use in all sorts of projects. You can use them in any soft circuit project that you don't want to be on all the time. How about hacking Project 8 so that the heart sparkles only when you squeeze it? You even could try making a big squishy pillow for your head that uses a little vibration motor to massage you when you lie on it.

PROJECT 10

Constellation Night Light

Star gaze and level up your DIY electronics skills

Tools

- One 8-inch embroidery hoop
- Sharp scissors
- Needle-nose pliers
- Two needles
- Sewing pins
- Ruler and chalk

Materials

- One 10-inch square of soft felt in navy
- Seven 3-mm white LEDs and one Teknikio star LED
- Sewable on/off switch
- White embroidery thread
- Conductive thread
- Two sewable 3V battery holders and batteries

As children, my sister and I had a battered and much-loved copy of a book that showed constellations of stars and revealed the myths and legends behind them. The stars we see at night change depending on where we live in the world and what time of year it is, but in my home country of Wales, there was one we could always rely on: the Plough (also known as the Big Dipper) and the North Star. This project lets me keep my favorite constellation shining in my home no matter where I am in the world.

To make your own constellation night light, you'll need to sew a circuit with eight LEDs. Working with this number of LEDs means that you'll have to learn how to link batteries so that you can add more power into your circuits.

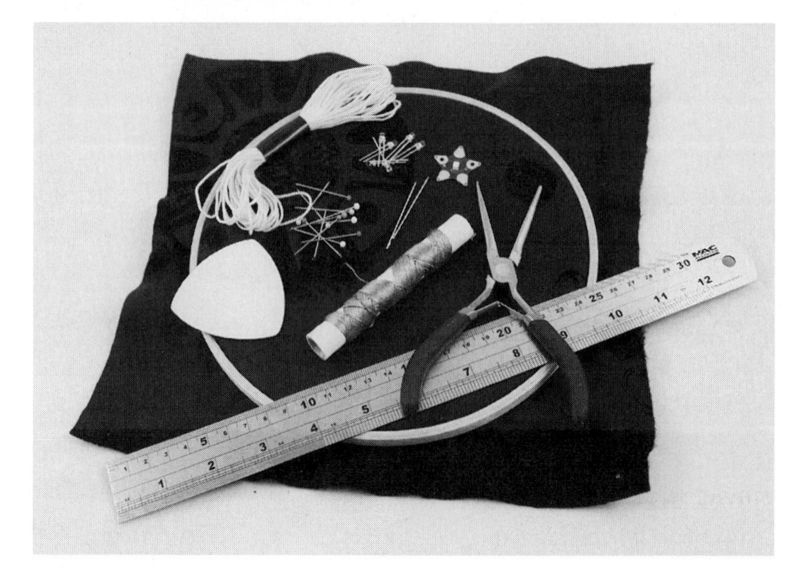

Preparing Your Materials

Make sure that you've got all your tools and materials ready. Then put your felt fabric into the embroidery hoop to keep your work secure while you sew. For this project, I like to leave the material inside the embroidery hoop when it's done. The hoop makes a very nice frame, and you can easily hang it on your wall with some thread looped around the tightening bolt.

Once your fabric is in place, mark your pattern. See page 250 for my template pattern of the Plough and North Star constellation night light. For this pattern, I use pins, chalk, and a ruler to mark my material.

Trace the constellation onto a piece of paper, and then place the paper over your fabric. Push sewing pins through the dots on the paper into the fabric. These will be the stars. Keeping the pins in place, lift one corner of your paper until you see a pin sticking through the fabric. From underneath the paper, push another sewing pin through the felt in the same place. Remove the original pin that came through the paper, and move on to the next star. At the end of your pinning, you should have eight pins and no paper. You can then use chalk and a ruler to draw a straight line between each of the stars. This will help you to neatly embroider the connections between your stars using a backstitch.

How to Embroider with a Backstitch

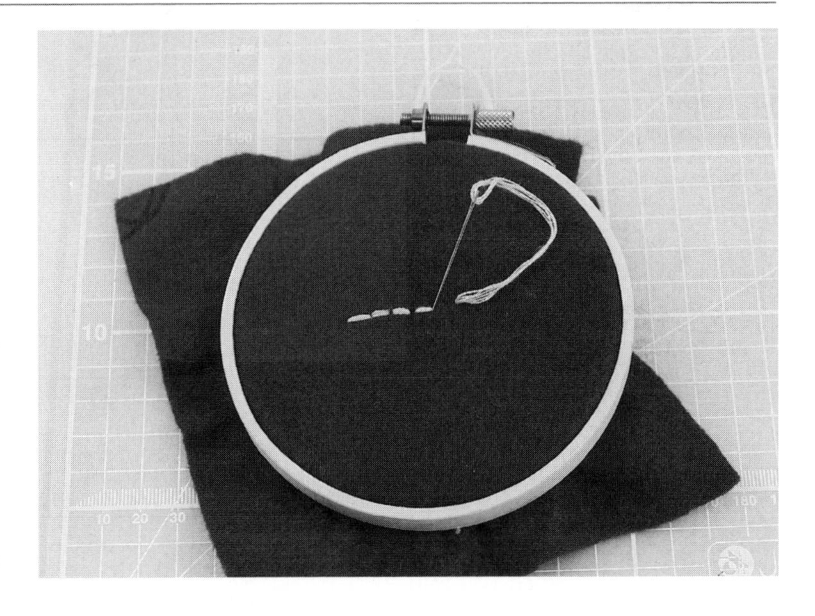

In this project, I like to do my embroidery stitches before I add my electronics. We've been using the basic running stitch a lot already in this part, but this time we'll be adding a new sewing technique and a new material: *backstitch and embroidery thread.*

Embroidery thread is made of six strands of cotton, which you'll need to separate out before using. This technique makes it hassle-free. Take the end of a single strand of thread, and hold it firmly. With your other hand, gently pull the rest of the strands away. Repeat this process to get more than one strand. Choosing how many strands to use in your project depends on the fabric you are using and the effect you are trying to achieve. Put all the strands you want to use together, and thread a needle as usual. If you are double threading your needle, this will double the amount of strands per stitch.

The backstitch is a very important stitch in embroidery. It is strong, neat, and makes attractive lines. It will take you a couple of goes to master it, but don't give up! Make a single, straight stitch about ¼ inch long. Continue along your pattern line from the underside of your fabric, and come back up to the surface of your fabric ¼ inch ahead. Next, bring your needle back down into or near the same hole at the end of the last stitch you made. This is why it is called a backstitch: you are bringing your stitch back to the end of the last stitch. If you're finding it tricky, search online for a video to help you.

Sewing Your Constellation Night Light

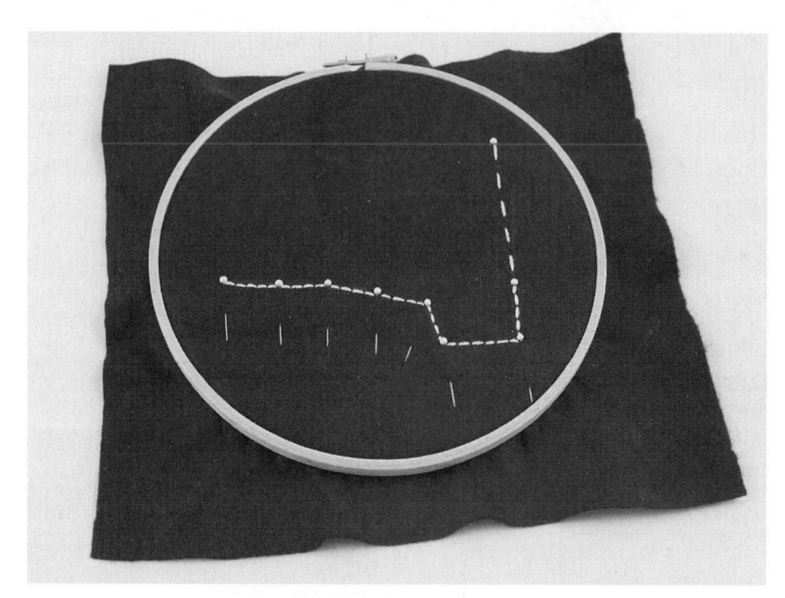

Once you've got the hang of the backstitch and have played with a few different ways of using embroidery thread, it's time to choose how to sew your constellation. Before I start a project, I spend some time trying different combinations of stitches and threads to practice getting the effect I want.

I decided to use three strands of embroidery thread with a double-threaded needle, meaning that I was sewing with six strands in total. I used ¼-inch backstitches for the connections between the stars in the Plough constellation and then a slightly bigger running stitch with equal gaps and stitches for the connection between the Plough and the North Star.

Once you are happy with your stitch choice and how many strands of embroidery thread you want to sew with, sew the connections between the stars onto your marked-up fabric.

At this point, you can remove your sewing pins, or if you'd like a little more guidance, you can leave them in until you add the LEDs.

Adding Your North Star

Take your Teknikio star, and place it where the North Star goes on your constellation night light. Thread a long piece of conductive thread onto your needle, tie it off, and then secure your knot. You'll need your thread to be about 15 inches long. We'll be using the same double-needle technique we used in Project 8, so thread a second needle with a piece of conductive thread.

Using one of your needles, securely sew the positive point of the Teknikio star to your fabric, and then sew on the negative point of the star using the other needle and thread.

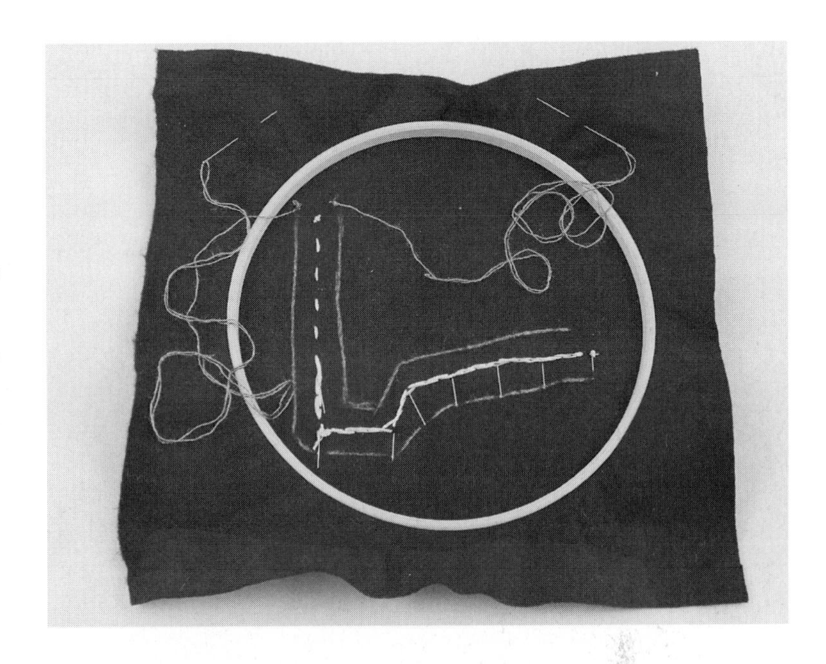

Check that your star is working by touching the needle connected to the positive leg to the positive side of a 3V battery at the same time as you touch the needle connected to the negative leg to the negative side of the battery. For the rest of this project, you should check that your circuit works after sewing every LED. This will save you lots of time unpicking mistakes later!

Now that your star is in place, flip your embroidery hoop over, and chalk a line to follow for the positive and negative paths of your electricity, just as in the picture. You'll notice that we're using a parallel circuit for this project. This means that we'll be connecting all the positives together and all the negatives together. Don't mix them up, or your circuit won't work!

Stitching in Your Stars

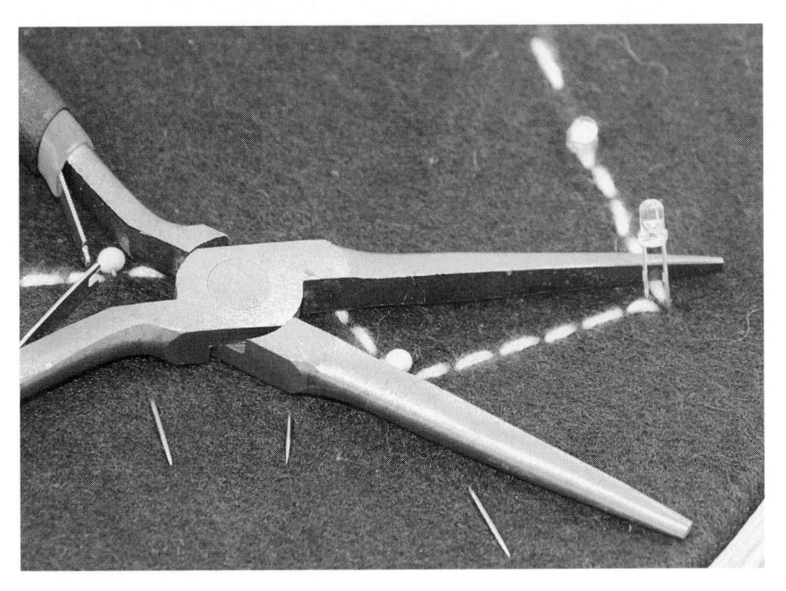

Next, we're going to add the LEDs. Follow this process for each LED one by one. When you add each LED to your constellation night light, make sure that you put your LED through the fabric the correct way around so that the positive and negative legs can be sewn to the correct path that you've marked out on the back of your project.

Use a needle to make a couple of holes, and poke the legs of the LEDs through the felt, just like you did in Project 8. Wiggle the needle round a bit to make a little hole, and then gently spread the legs of the

LED apart a little bit and push them through the holes you just made. Don't force them through or you might break the LED. Double-check that you've poked the LED through the correct way around: positive leg to the positive path and negative leg to the negative path.

Wrap each of the LED legs around the end of your needle-nose pliers into a circle so that you can sew it easily. Then gently flatten each leg against your material. Once you have your first LED in place and ready to sew, use the needle connected to the positive point of the star to sew the positive leg of the LED. Do the same for the negative side, and then test your circuit.

Finishing Your Constellation

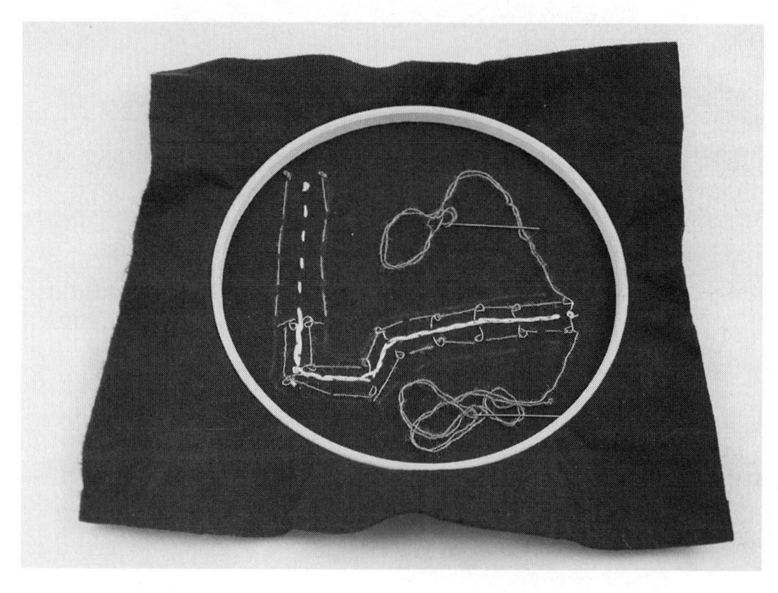

When I was sewing this project, I didn't want to show too much of my conductive thread on the front of my constellation night light. To minimize the amount of thread showing, I made bigger stitches on the back and teeny tiny stitches on the front.

I also used a thicker navy felt for this project so that I could stitch the LEDs onto the fabric without going all the way through to the front of the material. This trick can be pretty fiddly depending on the fabric you're using, so don't worry if you're not able to do it. The conductive thread looks nice anyway: just try to keep your stitches small and neat on the front part of your constellation night light.

However you choose to stitch your stars, repeat the last step for all seven stars in the Plough constellation. Push the LEDs through the fabric, and use your pliers to make them sewable one at a time. Take care that they go through the fabric correctly aligned with the positive and negative paths that you've marked out on the back of your project. After sewing both legs of each LED in place, make sure that your connections are secure, and test your circuit as you go. Check the "Fix It" box at the end of this project if you run into any problems.

Adding an On/Off Switch

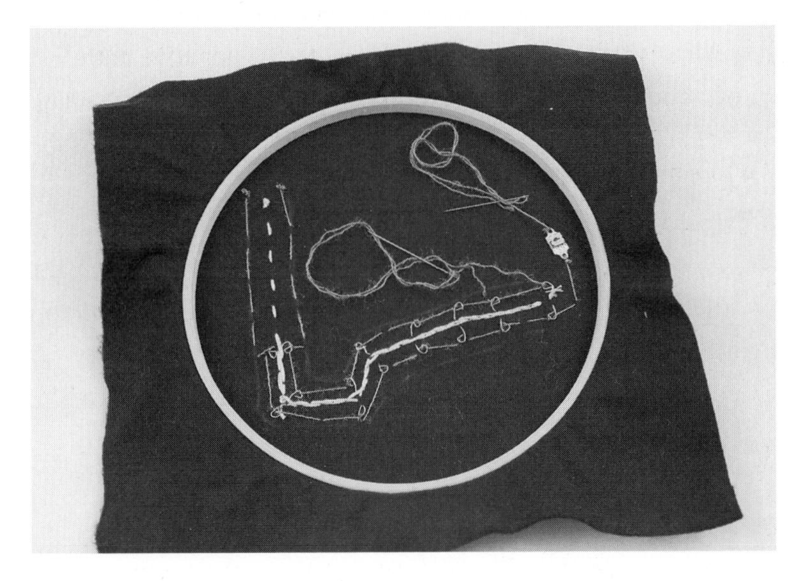

When you've finished making your constellation night light and have it hanging on your wall, you probably don't want to have it shining all the time. We're going to add a sewable on/off switch to our project to help us do that.

If you cut the thread of a completed circuit, it will stop working. This is because you'll have prevented the flow of electricity from completing its journey around the path. If you reconnect the ends of the thread, the power will start flowing again. Switches let us disconnect and reconnect the power in our circuit without having to cut the thread or take the batteries out. We made some basic switches in the Paper Circuit part, and we'll be learning more about them—and making our own DIY switches—in Project 11.

Turn your constellation night light over, and stitch one side of a sewable on/off switch to either the positive or negative path of your circuit. It doesn't matter which path you choose. After you've securely fastened one end of the sewable switch to the fabric with your conductive thread, tie off your thread, and secure your knot with nail varnish. Rethread your needle with conductive thread, and stitch the other side of the sewable on/off switch to your fabric. Test your circuit with the switch in the on and off positions.

How to Link Batteries Together

As we learned earlier, each type of battery has a different amount of power it is able to provide. The power in a battery is measured using volts. This is what the V in 3V battery means. We also know that each component uses power. When I design circuits, I have to think about the balance between the available power from the batteries, the amount of energy it takes to push power through my components, and what that means for the electricity flowing around my circuit.

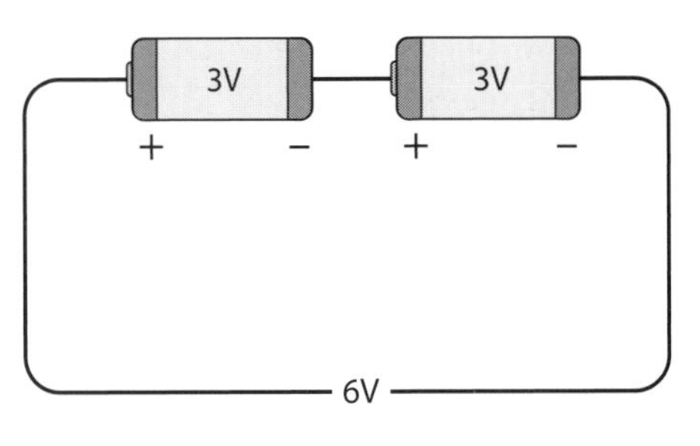

Batteries in Series

We will learn more about this balance later in this book, but for now let's talk about linking batteries. You can wire batteries together in series or in parallel in the same way that you can wire LEDs together. If you wire two batteries in series (negative to positive, like a daisy chain), you increase the voltage available. For example, linking two 3V battery packs in series will give you 6V. If you wire two battery packs together in parallel (positives linked and negatives linked), the voltages of both batteries are added together and divided by 2. For example, a 3V battery and a 9V battery wired in parallel will give you 6V.

The more voltage you give an LED, the brighter it shines. However, if you give your LEDs too much voltage, you will destroy them. Six volts of power is too much for our LEDs, so to make our constellation shine brightly for a longer time, we're going to link two 3V battery packs in parallel. Using the knowledge we just learned, can you work out many volts two 3V battery packs wired in parallel will provide?

Sewing Your Batteries

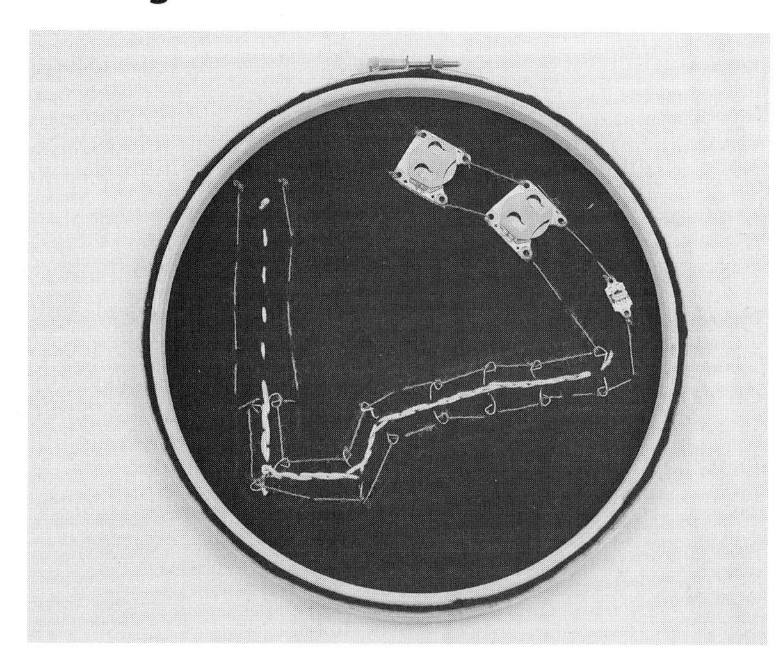

Use the conductive thread on your positive path to stitch the positive holes of a sewable 3V battery holder securely onto the fabric. Next, place a second 3V battery holder near the first, and connect its positive holes using your conductive thread. Tie this side of the thread off, and secure the knot in place with a dab of nail varnish.

Next, take your needle and conductive thread from the negative side of your parallel circuit. Stitch your way to the first 3V battery pack, and connect the negative holes with your thread. Sew your way to the second battery pack, and connect those negative holes too. Tie the thread off, and secure the knot.

Finally, add your batteries, and flip the switch. How does it look? Where are you going to hang your creation?

Fix It

Not working? Don't worry! Follow these steps to figure out why, and fix it.

1. Check your power.
   - Are your batteries the right way around? Flip them over and see what happens.
   - Have they run out of juice? Try new batteries.
   - Are your batteries connecting into your circuit? Make sure that the conductive thread is connecting to the correct sides of the battery packs.
2. Check your components.
   - Are your LEDs working? It's a good habit to check each component before you add it into a circuit, especially when you're using lots of components in a project such as this.
   - Are your LEDs securely sewn in place? Loose connections mean that your circuit won't work. Tighten your connections, and try again.
   - Is your switch turned on? It may seem obvious, but we've all made this simple mistake before!
3. Check your wiring.
   - Do you have a short circuit? If your positive and negative paths touch, no matter how slightly, your circuit won't work. Tidy up your loose ends, check your knots for fraying, and restitch any crossing paths. This project uses a lot of thread, so it's very easy for fraying thread on your path to short circuit your project. Use your finger to gently brush between the positive and negative paths, removing any stray strands.

Make More

This project is a great way to show off some more advanced embroidery techniques. Once you've made sure that your main stars light up, why not look up how to do a French knot? You can use this pretty embroidery stitch to embellish your constellation with silver or white dots to represent the other stars in the night sky.

You can also expand this project into a bigger collective piece. Why not get together with your friends to design and make a whole night sky map. The stars you can see vary at different times of year and in different locations. You could choose to represent your favorite place at your favorite time of year or show what the night sky looked like for one special occasion, for example, on the day you were born.

PROJECT 11

Grumpy Monster with DIY Tilt Sensor

Build a grumpy monster that complains when it falls over

Tools

- Sharp scissors
- Needle-nose pliers
- Three needles
- Sewing pins
- Tracing paper and a pencil

Materials

- One 10- × 10-inch square of felt
- One sewable buzzer
- Cotton thread
- Conductive thread
- One sewable 3V battery holder and battery
- Buttons, beads, pompoms, or other craft materials

In this project, you'll be designing and building your own unique squishy grumpy monster. Then you will add a sewable buzzer and make a special kind of switch that activates only when your monster is upside down. This kind of switch is called a *tilt sensor*. You can design a tilt sensor to complete a circuit only when it is orientated in a certain direction, making it ideal for projects that you want to activate with movement.

This project is also a great opportunity to use a variety of the sewing and craft skills you've learned in this part. You can show off your embroidery skills with running stitches, backstitches, and blanket stitches, *and* you'll get to embellish your monster with buttons, beads, pompoms, material cutouts, or anything else that you think will fit your monster's personality.

Preparing Your Materials

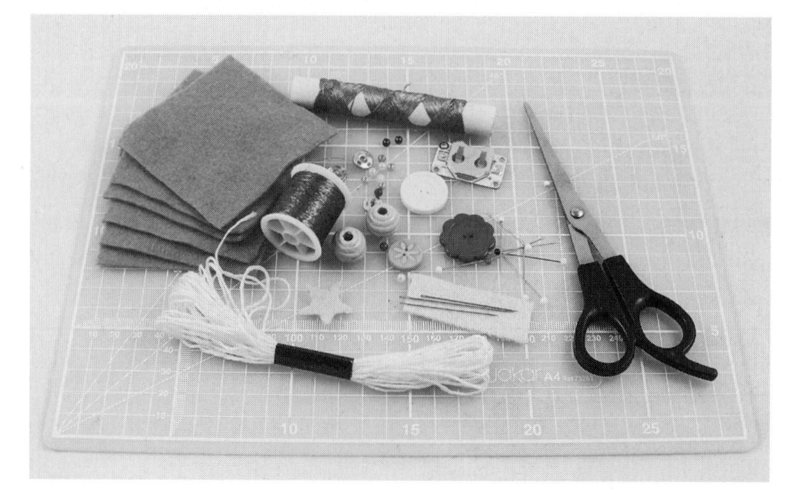

Make sure that you've got all your tools and materials ready. Then prepare your felt. Turn to page 253 for a template of the grumpy monster cube pattern. Trace and cut out the squares, and then pin them to your felt and use them as a guide to mark out your pattern. Cut all six squares, and check them against each other: they should be the same size.

For this project, you'll also need a selection of craft materials. I used cute mismatched buttons for eyes, pompoms for ears, beads for a tail, and felt for facial features, but you can design your monster in any way you like. Try raiding a button tin or visiting a thrift shop for inspiration.

You will need something conductive for the end of the tilt sensor tail. I used a snap-press stud, but you can use a button made of conductive metal, a jingly bell, or a conductive metal bead.

Designing Your Monster

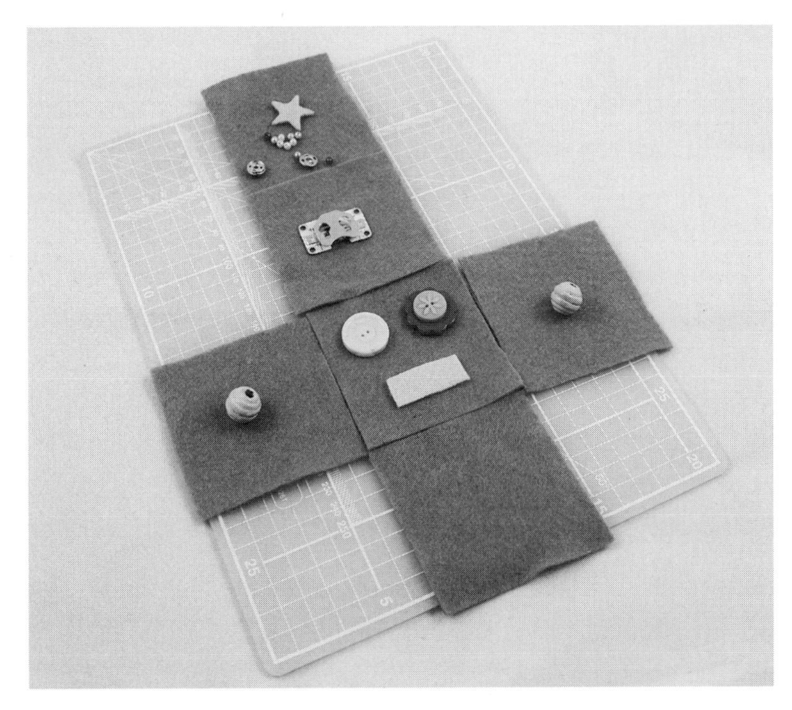

Each of the six squares forms one side of your cuboid grumpy monster. Lay them out as shown in the picture, and experiment with your embellishments and components until you're happy with your design.

The three cubes in a row toward the bottom of the picture are the face and sides. There are no electronic components on these three squares, so you just need to decide on a crafty design that fits your grumpy monster's personality. The two sides could be ears, arms, tentacles, wings, or anything that tickles your imagination.

The square below the face is where your monster normally sits, so if you do decide to embellish it, make sure that your design can lie flat. The square above the face is the top of your monster's head. It will need to have a battery holder sewn in the middle, so don't add too much other stuff. The square at the very top of the picture will be the back of your grumpy monster, where the tilt-switch tail goes. I've sewn a star on my monster's bottom for the tail to come out of, but you can leave it bare or add your own design. Just make sure that any additions are not bulky enough to get in the way of your switch.

Once you've laid out your design, sew all the embellishments in place. Make sure that you leave at least a ½-inch margin around all sides of the squares so that your decorations don't get in the way of the blanket stitch later on.

Sewing the Battery Pack

Double thread two needles with conductive thread, tie off, and secure your knots. Take the square that will become the top of your grumpy monster's head, and sew on the positive holes of the battery pack from the front of the head square toward the back. Once the positive side of the battery pack is in place, continue sewing toward the back of the square. Remember to use small, neat stitches and check the underside of the fabric for loops, bumps, and tangles. When you get to ¼ inch from the edge of the fabric, pause sewing this positive path, leaving the needle and the conductive thread attached to the square for later.

Next, sew the negative holes of battery pack to the fabric, again from the front of the head square toward the back. Once your negative holes are securely sewn on, make a negative path to the back of the head that stops ¼ inch from the edge of the fabric. You should now have two parallel paths going from the battery to the edge of the square.

Tuck the needles into the square to keep them out of the way while you start sewing the cube. Now that your design is stitched in place and you have started your circuit, it's time to start constructing your cube. The best way to do this is to use a blanket stitch, which you may have chosen to use in Project 9. If you haven't tried a blanket stitch before, it's worth practicing a couple of times before starting to sew up your monster.

Sewing Your Cube with Blanket Stitches

Take ¼ inch of material from the adjoining edge of each square, and pin the face and sides together so that you have a row three squares long. Thread your needle with ordinary thread or embroidery thread. I used two threads of embroidery thread (four threads after doubling) in a contrasting color to my material.

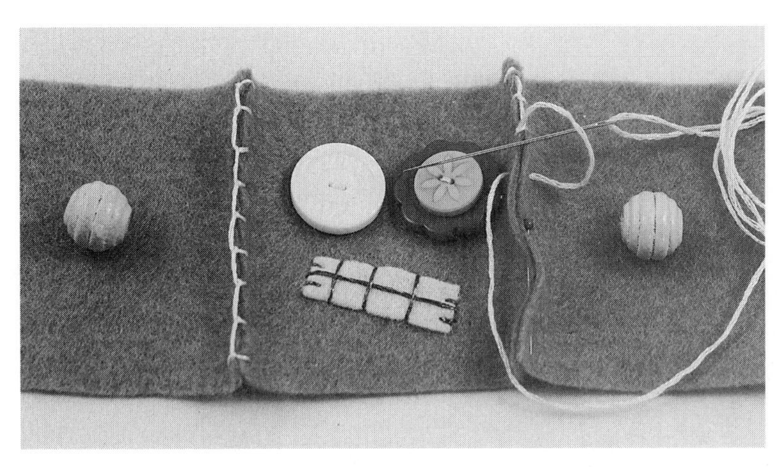

Start your blanket stitch at the top of one of your pinned sides, coming up from the inside of the square so that the first knot is hidden. Push your needle up through your fabric about ¼ inch from the edge. Pull the needle around to the front of the face, and then push it through to the same place you started, making a loop around the edges of the fabric squares. Before you pull the loop closed, send your needle from the top of the square toward the bottom, underneath the loop stitch you just made. Pull the thread through so that it is secure but not too tight.

Next, poke your needle from back to front about ¼ inch along to the left of your first stitch. As you pull that thread through, you'll see another loop. Again, push your needle through the loop from top to bottom under the stitch you just made, and then pull the thread through. That's your first blanket stitch!

If you're finding blanket stitch tricky, search online for a video to help you master this useful technique.

Constructing Your Monster

Keep sewing down the side of your monster's face, pulling out the pins as you get to them. To get the best results, keep the space between each stitch the same (about ¼ inch), and do the same for the distance of each stitch away from the edge (about ¼ inch). At the bottom of the square, tie off your thread. Then start again at the top of the other side of the face.

Once you've sewed the row of three squares (the face and the two sides of the head), you can do exactly the same thing with the bottom square. This time, attach one edge of the bottom square to the bottom edge of the face using blanket stitches. Next, add the top of your monster's head. This is the square where your circuit starts, and it should attach at the top of the face. Make sure that the paths of your circuit point toward the back of your monster.

Finally, add the back of the monster, where the tail will go. Carefully pin your tail square to the square with the battery pack. When the six squares are attached, they should look like the picture. Stitch the tail square in place, but make sure that you've tucked your needles and conductive thread to one side before you start sewing. Be super careful not to accidentally catch your conductive thread paths while sewing your blanket stitch. Once all six squares are attached together, you can start sewing your DIY tilt sensor.

What Is a Tilt Sensor?

A *tilt sensor* is a kind of switch that completes a circuit when it is facing in a certain direction. Many tilt sensors use mercury or balls made of conductive metal to make the connection, but I'm using conductive thread and press snaps. You can use anything conductive for your own sensor: metal buttons, a jingly bell, or a metal bead.

I've designed the circuit for this project to be open when our monster is the right way up. This means that our monster can't complain because there will be no electricity flowing through the circuit. However, when the monster is upside down, gravity means that the tail (which we've made out of conductive materials) will fall down and touch a tilt sensor contact pad. This closes the circuit, meaning that electricity can flow through the buzzer.

In the picture, you can see an example of a tilt sensor in an open circuit (where the tail is not completing the circuit) and in a closed circuit (where the tail is completing the circuit, allowing the electricity to flow).

That's how we use a tilt sensor to make our grumpy monster complain loudly about being upside down. Tilt sensors can help you do all sorts of cool projects, so this is a good DIY electronics technique to master.

Sewing Your Tilt Sensor Contact

Take the needle and conductive thread that is attached to the negative path, and sew a path from the battery pack square to the tail square. Remember that this will be a cube when you're finished, so don't sew your stitches extra tight because you'll need a little wiggle room. Don't make them too loose, though, or you could risk a short circuit!

Once your negative path is on the tail square, sew a path to the center of the square, about ½ inch from the top edge. Sew something conductive just there: this will form the contact of the tilt sensor. I've used a press stud, but you could use a metal bead, a button, or anything else that is conductive and a little raised up from the surface of the square.

Once the contact is securely in place, tie off your thread, and secure the knot. You are now ready to add the buzzer and finish your tilt sensor tail.

Sewing the Buzzer

Take the needle and conductive thread attached to the positive path, and sew from the battery pack square to the tail square. Again, make sure that your stitches are neither too loose nor too tight. Once the positive path is on the tail square, start sewing straight down until you're about ½ inch away from the top of the square. Now take your sewable buzzer and sew the positive hole onto the underside of your fabric. Once it is in place, tie off your thread, and secure the knot.

Thread another piece of conductive thread, and secure the knot. Stitch the negative hole of the buzzer onto the underside of your fabric, and then continue stitching, taking your path to the very center of the tail square. Push your needle through to the front side of the tail square, and you're ready to finish sewing your DIY tilt sensor switch.

Making Your Tilt Sensor Switch and Finishing Your Monster

Measure the distance between the center of the tail and the tilt sensor contact. This is how long you want your tail to be. Thread a few small beads onto your conductive thread until your tail is almost long enough to touch the tilt sensor contact.

Now stitch the conductive item you've chosen to make into the end of your tail. If you're using a bead, go around and through the center of the bead a few times to make sure that you've got a good contact. Finally, tie your thread off, and secure the knot. Add a battery, and test that your circuit works. Then finish off your monster.

Use pins and blanket stitches to sew the flat shape into a cube, one side at a time. When you're on the last open side, stitch about halfway along the edge, and then stuff your grumpy monster with wadding to help it keep its shape.

Finish your stitching, tie the final knot, and enjoy playing with your grumpy little monster.

Fix It

Not working? Don't worry! Follow these steps to figure out why, and fix it.

1. Check your power.
   - Is your battery the right way around? Flip it over and see what happens.
   - Has it run out of juice? Try another battery.

- Is your battery connecting into your circuit? Make sure that your battery fits snugly into the holder and that the conductive thread is connecting to the correct side of the battery.

2. Check your components.
   - Is your buzzer working? It's a good habit to check each component before you add it into a circuit.
   - Is your buzzer securely sewn in place? Loose connections mean that your circuit won't work. Tighten your connections, and try again.
3. Check your wiring.
   - Do you have a short circuit? If your positive and negative paths touch, no matter how slightly, your circuit won't work. Tidy up your loose ends, check your knots and thread for fraying, and restitch any crossing paths.

Make More

Why not make your grumpy little monster some friends to keep it company? Now that you know how easy it is to use blanket stitches to make a three-dimensional cube, try making bigger and smaller cubes using the same technique. You could even experiment with other shapes. Could you make a plush pyramid?

In addition to making your designs more shapely, this project has taught you how to make a DIY tilt sensor. Try adding tilt sensors to another project. How about a copper tape tilt sensor that beeps when the lid of a cookie jar is opened or a light-up card that turns on only when it is opened?

Maker Spotlight

Name: Rachel Freire
Location: Mostly London
Home: Liverpool
Job: Artist and technician

Rachel works in fashion, costume, and garment technology. She loves mixing up futuristic and traditional techniques. Rachel launched her fashion label in 2009 at London Fashion Week, and she now works with designers, technicians, and engineers to make sustainable futuristic fashion. She is currently exploring the possibilities of e-textiles, modular fashion, and open hardware. She is the textile designer of MI.MU Gloves, an instrument that lets you compose and perform music with the gestures of your hands. Rachel also designs and makes costumes for Hollywood films, including *Guardians of the Galaxy*, *Avengers: Age of Ultron*, and *Maleficent*.

Q. ***What do you like to make and craft?***
A. I make all kinds of clothing: anything that lives on your body. I know from my background in costume design that clothing can be very powerful, creating characters and telling stories. I like to work with practical materials, so my favorite fabrics are stretchy and can be worn in many situations. Mostly now I work with e-textiles, which means fabric that conducts electricity, replacing wires and sensors. The things I make are halfway between clothing and costumes. This field is known as *wearable technology* or sometimes *fashiontech.*

Q. ***What are you currently working on?***
A. I am currently designing a new product called MI.MU Gloves so that we can all make music with just our hands. They are WiFi gloves filled with sensors that you program so that each movement and gesture makes a different sound. They can be used by musicians, performers, or dancers. They can also be used to control lights and projections. In the future, they could be used to control robots, games in virtual reality (VR), or even things in space.

Q. ***What is your dream project?***
A. I would like to have a fashion label for clothes that are completely customized for the person who buys them. We would design basic garments, and each person could

then add different colors, shapes, and functions. These designs would be completely open source so that everyone can be inspired by each other's ideas. In the future, we need clothes that last longer and are designed by each person so that they are unique.

Q. ***Who inspires you?***
A. I'm inspired by stories and storytellers. I love working on movies because they build huge new worlds. When you work on movies like *Guardians of the Galaxy*, you use new and unusual materials and techniques, which inspires other projects, such as my technology projects. This means using science fiction to make real-life garments for the future! Making costumes for big films like *Marvel* means that you are part of telling a huge story that inspires millions of people. If we tell good stories, we will make good futures.

Q. ***Can you share your favorite books, websites, or places you go to learn new skills?***
A. My absolute favorite place is KOBAKANT: How To Get What You Want (www.kobakant.at/DIY). Mika and Hannah do lots of experiments with different materials and document them all on this website. I have been using the Chibitronics book *Learn To Code* to teach my nephew about electricity and circuits. I also love the book *Getting Started in Electronics*, by Forrest M. Mimms III, because it is all handwritten and has great illustrations.

Q. ***What is your favorite place to get materials or tools?***
A. I like to go to my local fabric stores just to be inspired by random things I find. I take a multimeter so that I can check if any of the fabrics are conductive! I also use www.lessemf.com for conductive fabrics, Adafruit for electronics, and I love searching eBay for vintage tools.

Website: www.rachelfreire.com
Instagram: @rachelfreirestudio
Twitter: @rachelfreire
Work in progress: www.flickr.com/photos/rachelfreirestudio/albums

e-Textile Crystallography

I collaborated with Melissa Coleman for this project, which is all about growing crystals on LEDs using a special kind of salt. We made a fabric circuit and stitched on the LEDs. Next, we sealed all the electrical connections with clear nail varnish to make them waterproof. Then we dipped the whole circuit in a bath of salt and watched the crystals grow on the surface! You can mix the salt in different ways to make the crystals grow quickly or slowly. When they grow quickly, they are not as strong and can break

easily. You can also color them with dye or food coloring to make different colors. In each experiment, we make a larger circuit, and eventually it will become large enough for a whole dress.

Website: rachelfreire.com/etextile-crystallography/instructables.com/id/Growing-Crystals-on-LEDs-and-ETextiles/

Stretchy Circuits

When we think of electronics, we think of wires and boxes. This project is about making circuits that work with our bodies and our movements. You can make stretchy circuits using stretchy materials or by stitching in zigzags. I use conductive thread stitched in zigzags, allowing them to stretch, or stretchy silver-coated Lycra fabrics. I bond these to other materials with special stretchy glue. These stretchy circuits can be used to light up LEDs or to make stretch sensors that can measure how much your body moves. In the picture you see a girl wearing an outfit that has lots of lines. These are conductive fabrics that move electricity around the body. The black lines on her gloves are stretch sensors, and they can measure how much her fingers bend. In the future, the circuits in our clothes will become completely invisible!

Website: rachelfreire.com/second-skin, www.instructables.com/id/Stretch-Circuit/

PART THREE

Wearables

Introduction to Wearables

Welcome to the Wearables part. *Wearables* are electric circuits made for wearing on or near our bodies. They can be made using e-textiles, soft circuits, flexible circuits, or microcontrollers with bendy silicone wires sewn into clothing or accessories. These days, many new technologies are designed to be wearable, including musical instruments in gloves, devices worn on your spine to help you correct your posture, fitness-measuring devices worn on your heart or head, and motion-capture bodysuits used in Hollywood for making fantastical characters look real on screen.

This part will help you take your first steps in the world of wearables. By the time you finish making all the projects in this part, you will have enough skills and knowledge to start making your own wearable projects. We've made all of our projects from scratch so far, but in this part we are going to start *hacking*. Hardware hacking is when you take something that already exists and change it to improve its performance or make it do something for which it was not designed. Hacking is a really fun way to mix up your DIY electronics.

Now that you know the basics of how electricity works and how to make circuits with paper and thread, we can start getting a bit more ambitious with our makes. You will learn more about *debugging* our circuits—that is, the process of figuring out what is going wrong—and how to use conductive fabric. Each of these new skills is used in one of the five projects in this part, but if you want to skip ahead or look back, here's where to find them:

Wearables: Essential Skills

- How to test for conductivity (pages 143–144)
- How to choose and use conductive fabric (page 166)
- How to avoid short circuits (pages 187–188)

We will also keep exploring new components and new concepts in this part. You will learn more about conductivity and how to use a continuity sensor to test your circuits. You'll also learn what light sensors and RGB LEDs are and then put your new knowledge into practice by using them in your own projects. Here's where to look up this information:

- How to power different colors of LED (page 145)
- How to use a conductivity sensor (page 148)
- How touchscreens work (pages 164–165)
- What is a light sensor? (page 176)
- What are RGB LEDs? (page 185)

On the next page, you'll find a recap of some of the most important skills we learned in the Soft Circuits section plus some new guidance to help you make your wearables. This part will introduce you to some new craft techniques that will be very useful when you make beautiful wearable things, including two knots commonly used in jewelry and an easy way to make impressive paper flowers. Here's where to look up information on these skills:

- How to make paper flowers (pages 155–156)
- How to make a lark's head knot (page 175)
- How to tie an adjustable sliding knot (page 179)

Wearables: Essential Skills

How to Make Good Connections

When you sew circuits, there are two major issues that you should watch out for. The first issue is loose connections. Sewable components have handy holes that make it much easier to sew with electronics. When you're using these holes to sew your component onto your fabric, be sure to make enough stitches to keep the component secure and the conductive thread firmly in contact with the component. Remember, if your stitches are loose, the electricity will not be able to get to the component, and your circuit will not work!

How to Avoid Short Circuits

Another problem you could have when sewing circuits is that you might cross your positive and negative paths. Electricity likes to take the easiest path, so if you touch the two paths together, the electricity will take a shortcut, meaning that your components won't get any power. When using conductive thread, you also need to watch out for loose ends. If your conductive thread is frayed, a wayward strand could short circuit your whole project. Try not to put the positive and negative paths too close to each other, and check the underside of your fabric while you're sewing for rogue loops, untidy knots, and loose stitches that could sabotage your circuit.

How to Wire Your Circuits in Different Ways

When you make a project with more than one component, there are two ways to do it. The first way is to wire the components in series, and the second is to wire them in parallel. These two different ways of wiring circuits have different effects. Each component (e.g., an LED or a vibration motor) uses a certain amount of power, and each type of battery (e.g., 3V or AA) has a different amount of power it is able to provide.

When you wire two components in series, the circuit divides the total power supply between the two components. When you wire two components in parallel, each component will receive the same amount of power. In practical terms, a circuit with two components and one battery wired in parallel will make your components shine more brightly than the same circuit wired in series, but it will also drain your batteries more quickly.

How to Make a Series Circuit

Components wired in series are connected end to end, meaning that the negative bit of the first component connects to the positive bit of the second component and the negative bit of the second component connects to the positive bit of the third component and so on. This whole daisy chain of LEDs should be connected as usual to your power supply: negative to negative and positive to positive.

How to Make a Parallel Circuit

Components wired in parallel use one wire (or thread) to connect all the positive bits of the components to the positive bit of your battery pack and another wire to connect all the negative bits of the components to the negative bit of the power supply.

Quick Start Tips

You'll find lots of good tips and tricks for making your wearables as you read through this part. Here are the most important things to remember:

- *When sewing with conductive thread, keep your stitches small, neat, and close together.*
- *Make sure that your components are sewed onto your fabric securely.*
- *Watch out for loose ends, fraying thread, rogue loops, or loose stitches.*
- *Conductive thread can be tricky to keep knotted. Secure your knots in place with a dab of clear nail varnish.*
- *Practice new stitches and techniques on scrap material before using them in your projects.*

Tools and Materials List

To make all the projects in this part, you'll need the following tools:

- Flexible tape measure
- Wire cutters
- One needle
- Sharp scissors
- Fabric scissors
- Craft knife
- Pencil and paper

You'll also need the following materials:

From a Craft Store

- Five 8-inch squares of felt in any color
- Ordinary thread
- Two pairs of metal press studs
- Crepe or tissue paper in any color
- Paper- or plastic-covered bendable craft wire
- Leather or nylon cord
- One sewable badge pin

From an Electronics Store

- Conductive thread
- Five sewable 3V battery holders, at least two with an on/off switch
- Six 3V batteries
- Three sewable Lilypad LED sequins in different colors
- Six sewable LEDs
- One square of conductive fabric
- One Teknikio sewable light sensor
- Three sewable on/off switches

From a General Store

- Clear nail varnish
- One headband with a material outer layer
- Two pairs of woolen gloves

PROJECT 12

Conductivity-Sensing Bracelet

Make a beautiful wrist cuff that doubles as a DIY electronics tool for fixing circuits

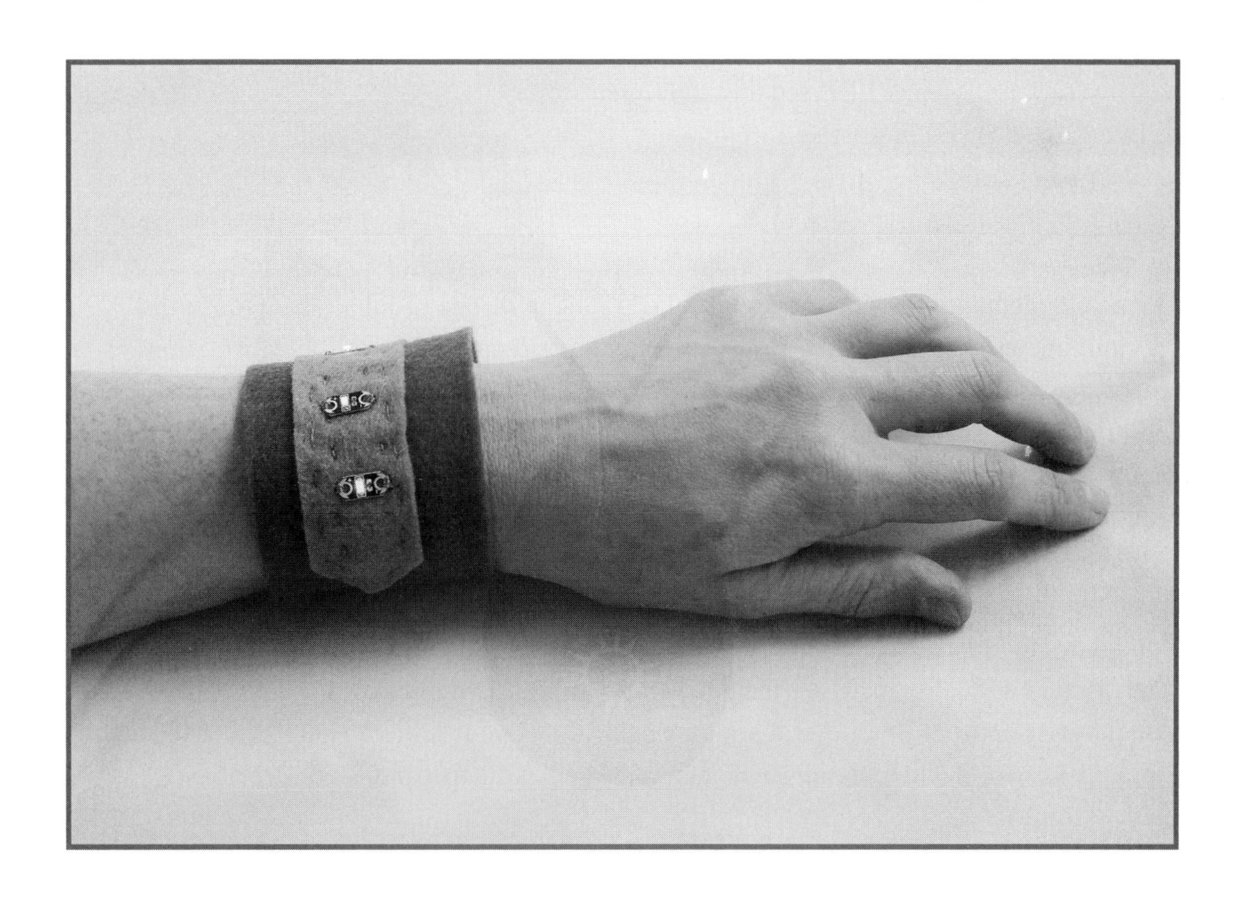

Tools

- Flexible tape measure
- One needle
- Sharp scissors
- Fabric scissors
- Pencil and paper

Materials

- One 8-inch square of felt in any color
- Conductive thread
- Ordinary thread
- One sewable Lilypad 3V battery holder and battery
- Three sewable Lilypad LED sequins in different colors
- Two pairs of metal press studs
- Clear nail varnish

The Conductivity-Sensing Bracelet is a DIY version of one of my favorite tools: the multimeter. A multimeter helps us to debug our circuits, giving us the information we need to figure out what is going wrong when we make a mistake. You can use your bracelet's conductivity-sensing powers to help you check that your copper tape or conductive thread circuits are connected where they should be and that they are not touching where they shouldn't be.

When we say that something is *conductive*, we mean that electricity can pass through it. When we say something is *nonconductive*, we mean that it will block the path of electricity. Conductive materials include copper tape, conductive threads, and some types of fabric, whereas nonconductive materials include paper, card, ordinary sticky tape, and plastic. Your Conductivity-Sensing Bracelet will not only help you fix your projects, but it will also let you explore your world and find out which materials are conductive and which are not.

Preparing Your Materials

Make sure that you've got all your tools and materials ready. Then prepare your felt. You'll need to measure your wrist to find the correct dimensions for your bracelet.

Use a flexible tape measure to find the circumference of your wrist, and then take that measurement and add 1½ inches. This final measurement should be the length of your bracelet. The width should be around 2 inches.

If you don't have a flexible tape measure, you can improvise by measuring your wrist using a piece of string and then using an ordinary ruler to measure the string.

Mark out your final measurements on your felt square using chalk or a water-soluble pen. Then use your fabric scissors to cut out the rectangle.

How to Test for Conductivity

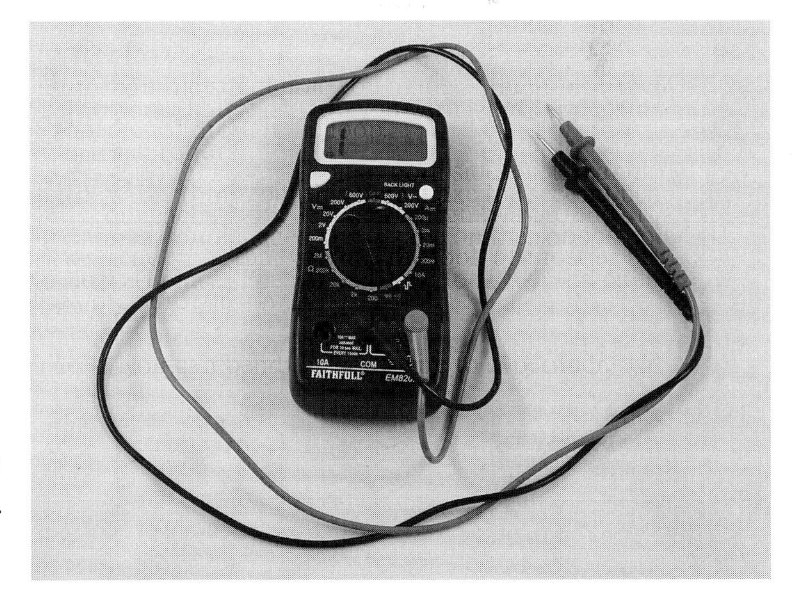

A multimeter is one of the most important tools in the DIY electronics world. It helps us figure out what is going wrong in our circuits by measuring lots of useful things: how much power is in our batteries, how much electricity is flowing around our circuit, and how much effort electricity has to make to get through a component.

A multimeter also has a function called a *continuity tester*, which checks whether different parts of a circuit are connected. Our Conductivity-Sensing Bracelet is a DIY version of a multimeter's continuity tester.

When you test whether two parts of a circuit are connected, you are testing whether electricity can move between those two points. For electricity to move between two points, they need to be connected with something conductive such as copper tape, conductive thread, or a wire.

If you find that there is not a connection between two points that should be connected, you know where the problem is, so you can fix it! In the same way, if there is a connection between two points that should not be connected, you know that you have a short circuit and your circuit won't work. Finding the problem is the first step to fixing it!

Sketching Your Circuit

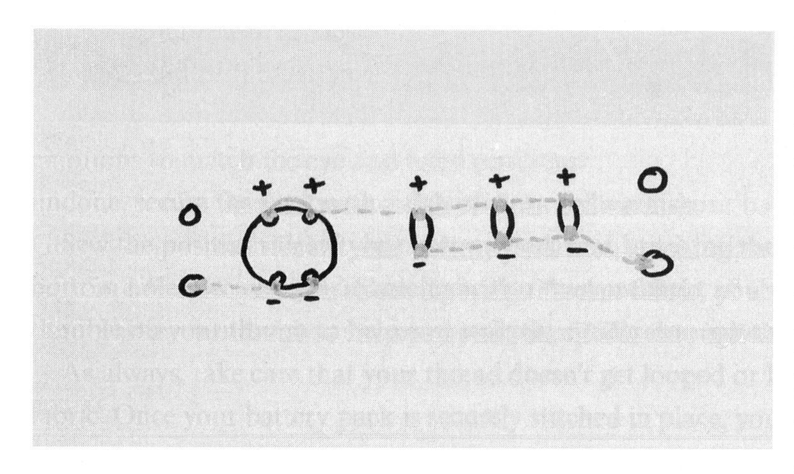

This project uses a component that we've not used before: LED sequins. They're similar to the sewable LEDs you've used before, but these LEDs are very small, and they come in lots of different colors. Try lighting them up individually to find your favorite colors, and then choose three to include in your Conductivity-Sensing Bracelet.

Once you've chosen your LED sequins, lay them out alongside your sewable battery pack and your press studs, just as in the picture. We're going to make a parallel circuit that is only completed when the bracelet is being worn or when the two press studs are touched to a connected conductive surface. Then the two press studs will connect, completing the circuit and allowing the electricity to flow.

Remember, LEDs wired in parallel use one wire (or thread) to connect all the positive bits of the LEDs to the positive bit of your battery pack and another wire to connect all the negative bits of the LEDs to the negative bit of the power supply.

Draw your components on a piece of paper, and then add in the circuit paths, just as in the picture.

How to Power Different Colors of LEDs

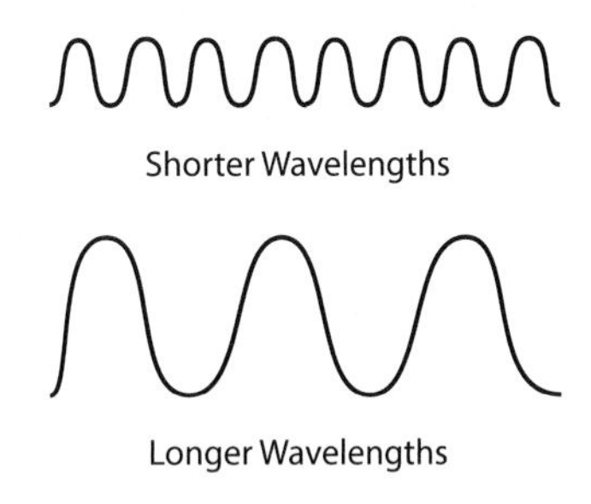

Wavelengths

So far in this book we've used white Chibitronics LED stickers and ordinary LEDs in white and yellow. In this project, we are using Lilypad or Adafruit LED sequins, which come in yellow, white, red, blue, green, and pink.

You will have noticed that the LEDs we use have different levels of brightness depending on the way that we connect up the batteries, the way that we connect up the components, and how drained our battery is. You may not have noticed that the brightness can also vary based on the color of LED you use!

This is because of a cool bit of science that I'd like to share with you. Light travels as waves, like ripples in a tank of water. Different colors of light have different wavelengths, like different sizes of water ripples. To produce light with a short wavelength, you need more energy.

Red light has a longer wavelength than yellow light, followed by green, blue, and finally, white, which has the shortest wavelength. This means that if you give a white LED and a red LED the same power in the same circuit, the red LED will be brighter. Why not experiment with some different colors of LEDs to see whether you can notice the difference? Use the same battery and connections in all your experiments to make the tests fair.

Designing Your Conductivity-Sensing Bracelet

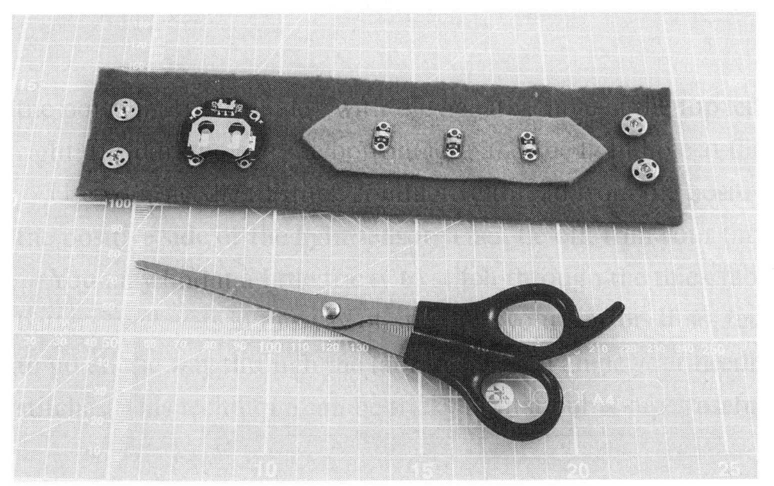

Now that you know what your circuit is going to look like and what it is going to do, you can start planning the layout and design of your Conductivity-Sensing Bracelet. You can sketch this as usual, but I designed my bracelet a little differently this time.

I knew that I wanted my bracelet to look simple and geometric, so I cut out a few different shapes and experimented with laying them out on the rectangular base. Once I had a combination of shapes and colors that I liked the look of, I arranged the components on top and took a photo to help me remember the layout when I start sewing.

Remember to lay out your components so that they match the circuit you designed. Because we're making a parallel circuit, this means connecting all the positive bits of your components and all the negative bits of your components.

Finally, remove the electronic components and the press studs and use ordinary thread to sew all your design elements in place.

Sewing Your Conductivity-Sensing Bracelet

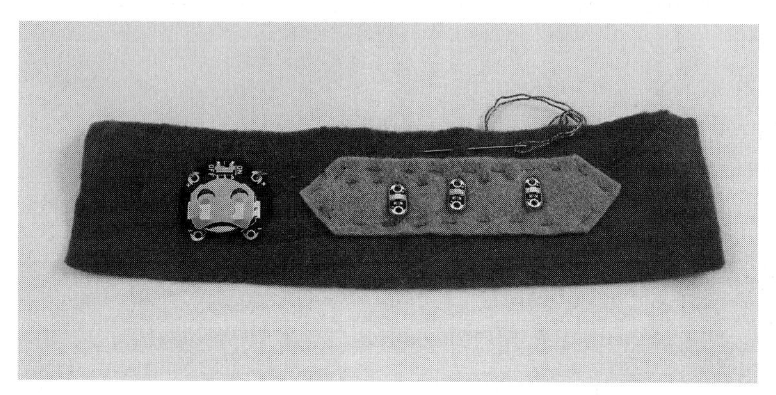

Thread a needle with a length of conductive thread, and tie a knot at the end. If your thread is the slippery kind that tends to come undone, secure your knot with a dab of clear nail varnish.

Sew the positive side of your battery pack first, stitching the left hole first (closest to the edge) and then the right hole (closest to the LEDs). Take care that your thread doesn't get looped or knotted underneath your fabric. Once your battery pack is securely stitched in place, sew a path over to the positive bit of your first LED.

Secure the positive bit of each LED in place one at a time. Make at least four stitches through each of the positive sewable holes to ensure that you have a good connection.

Once the positive bits of all three LED sequins are sewn in place, tie a knot in your conductive thread, cut the excess thread, and secure the end with a dab of nail varnish.

Sewing the Negative Side of Your Bracelet

Thread a needle with another length of conductive thread, and then tie and secure your knot. Start sewing the negative path from the first LED sequin on the left, not the battery pack. Stitch the negative bits of each LED in place one at a time, making sure that you have a secure connection and no fraying, looping, or knotting on the underside.

Once the negative bits of all three LED sequins are sewn in place, stitch across and down toward the place where the press stud will be located. Use conductive thread to stitch one half of the press stud to the bottom-right corner. Take a moment to make sure that you're sewing the press stud the right way up so that the other half fits.

Make sure that you sew over each of the sections of the press stud at least twice so that you make a good connection. The press studs get the most wear and tear once this project is finished, so you need to make your stitches secure. Once your press stud is sewn in place, tie a knot in your conductive thread, cut the excess thread, and secure it with a dab of nail varnish.

Finishing Your Conductivity-Sensing Bracelet

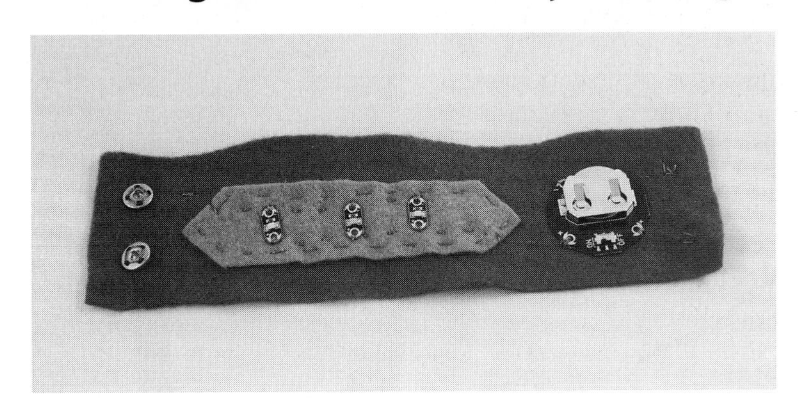

Now you need to connect the negative side of the battery to the press stud on the other side of the bracelet. Thread a needle with another length of conductive thread, and then tie and secure your knot. Sew the negative side of your battery pack, stitching the right hole first (closest to the LEDs) and then the left hole (closest to the edge).

Next, stitch across and down toward the place where the second half of your press stud will be located. Remember that your bracelet will be going around your wrist and that the two press studs will need to connect with each other. This means that the other half of your press stud will need to go on the underside of your fabric. Once you've double-checked the positioning of your press stud, use your conductive thread to securely stitch it to the underside of the bottom-left corner. Tie a knot in your conductive thread, cut the excess thread, and secure the end with a dab of nail varnish.

Finally, use ordinary thread to stitch the other two press studs in place. You've finished your Conductivity-Sensing Bracelet! Now you can use it to explore conductive materials, test that your paths are correctly connected, and check that you have no short circuits.

How to Use a Conductivity Sensor

You can use your Conductivity-Sensing Bracelet to test whether a material is conductive or nonconductive. Make sure that your battery is switched on, and then simply touch the two press studs that form part of your circuit to a material. If the material is conductive, your LED sequins will turn on.

You can also use your Conductivity-Sensing Bracelet as a tool for checking that your copper tape or conductive thread circuits are connected where they should be and that they are not touching where they shouldn't be. This works in the same way as when you were checking to see whether a material was conductive.

Switch your battery on, and then touch your press studs to two points in your circuit. If those two points are connected, your LED sequins will turn on. If they should be connected and your sequins do not turn on, you know where your problem is! In the same way, if they do turn on and they shouldn't, you've found your problem. Time to rewire!

Fix It

Not working? Don't worry! Follow these steps to figure out why, and fix it.

1. Check your power.
   - Is your battery the right way around? Flip it over and see what happens.
   - Has it run out of juice? Try another battery.
   - Is your battery connecting into your circuit? Make sure that your battery fits snugly into the holder and that the conductive thread is connecting to the correct side of the battery.
2. Check your components.
   - Are your LEDs working? It's a good habit to check each component before you add it into a circuit.
   - Are your LEDs securely sewn in place? Loose connections mean that your circuit won't work. Tighten your connections, and try again.

- Are your LEDs sewn the right way around? All the positive bits of the LEDs should be connected to the positive bit of your battery pack, and all the negative bits of the LEDs should be connected to the negative bit of the power supply via the press studs.

3. Check your wiring.
   - Do you have a short circuit? If your positive and negative paths touch, no matter how slightly, your circuit won't work. Tidy up your loose ends, check your knots and thread for fraying, and restitch any crossing paths.

Make More

If you like the idea of wearing your tools, you can make more things to keep your DIY electronics kit handy at all times. You can sew magnets into a wrist cuff to keep screws, nuts, and bolts from going astray, or you could make a tool belt by hacking an ordinary kitchen apron and adding custom pockets, loops, or hooks to keep your tools and components organized and ready to use.

PROJECT 13

LED Paper Flower Crown

Hack a headband into a stylish party piece with DIY paper flowers and LEDs

Tools

- Wire cutters
- One needle
- Sharp scissors

Materials

- Crepe or tissue paper in any color
- One headband with a material outer layer
- Paper- or plastic-covered bendable craft wire
- Conductive thread
- One sewable 3V battery holder and battery
- Three sewable LEDs
- Clear nail varnish

We've made all our projects from scratch so far, but a really fun way to mix up your DIY electronics is to hack things that you buy, find, or already own. *Hardware hacking* is when you take something that already exists and change it to improve its performance or make it do something for which it was not designed.

Some of my favorite makes of all time include parts that started life as something completely different. In this project, we're hacking a humble headband and making it into something really special. Keep an eye out for other things that you can fix, change, up-cycle, or reboot using your imagination, craft techniques, and DIY electronics skills.

Preparing Your Materials

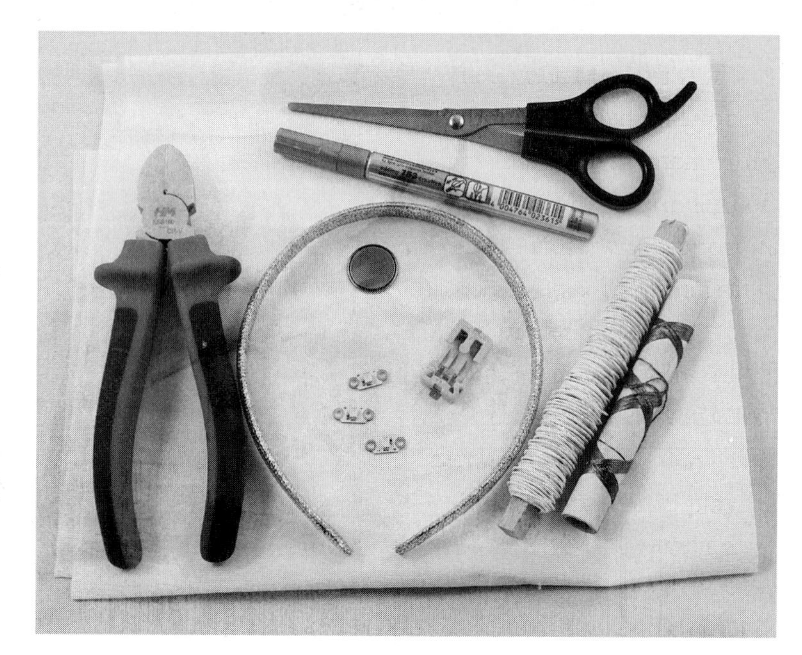

Make sure that you've got all your tools and materials ready before you begin. For this project, you'll need to find a headband that has a layer of material wrapped around a plastic middle bit. If your headband has metal in the middle, this hack won't work. These kinds of headbands are very common, so you should be able to find one at any clothing or accessory store.

I've chosen a thin headband for my hack, but if you'd prefer to use a thicker headband to make the sewing easier (or just because you prefer the look), then go right ahead.

You'll also need to find some bendable wire that is coated in paper or plastic. This kind of wire is easy to find in craft stores, online, or even at a florist. The reason we need to use wire coated in paper or plastic is so that the metal of the wire doesn't interfere with your circuit.

Hacking Your Headband

This hack uses one 3V battery and three LEDs. As you know by now, we'll need to make ourselves a parallel circuit to power multiple LEDs from one 3V battery.

Thread a needle with a length of conductive thread, and tie a knot at the end. If your thread is the slippery kind that tends to come undone, secure your knot with a dab of clear nail varnish.

Start your hack by sewing the negative path first. Securely attach your sewable battery holder to one end of the headband using the thread, with the negative hole facing down. Once your battery pack is in place, take your needle over to the underside of the headband—the side that would normally touch your hair.

On the underside of the headband, make a path of small, neat running stitches through the bottom layer of fabric until you reach the place where you want to put your first LED. Bring your needle around to the top of the headband, and stitch the negative hole of your first LED in place. Once it is in place, bring your needle back around to the underside of the headband, just as in the picture.

Completing Your Negative Path

Now that your first LED is in place, continue sewing on the underside of the headband until you reach the place where you want to put your second LED. Pause for a moment to look at the first LED that you sewed onto your headband.

When we sew the positive path, we need to have a clear run for our stitches, so make sure that you come up to the top of the headband the same way around as you did for the first LED. This will mean that your negative stitches are on the same side, making your job easier later.

Now bring your needle around to the top of the headband, and stitch the negative hole of your second LED in place. Once it is in place, bring your needle back around to the underside of the headband, and then repeat the whole process for your final LED.

Once you have securely sewn the final LED on the negative side, tie a knot in your conductive thread, cut the excess thread, and secure it with a dab of nail varnish.

Sewing Your Positive Path

Thread a needle with another length of conductive thread, and tie a knot at the end. Securely sew the positive side of your sewable battery holder in place, and then make a path of small, neat running stitches through the top layer of the fabric on your headband until you get close to your first LED.

Taking care not to cross your positive and negative paths, sew around the LED using the opposite edge to the side with your negative path, just as in the picture. Secure the positive side of your LED in place, and then continue along the top of your headband to sew the positive bits of the other two LEDs in place.

Once you have securely sewn the final LED on the positive side, tie a knot in your conductive thread, cut the excess thread, and secure it with a dab of nail varnish. Test that your circuit works by adding your battery.

Because my headband is thin, my positive and negative paths are quite close to each other. To keep fraying thread from breaking the circuit, I often use hairspray to keep my conductive thread strands in place once they are sewn. I give my paths a light spray and smooth the thread down gently with my fingers. This little tip is super useful in projects such as this with paths that are close to each other.

Preparing Your Paper Flowers

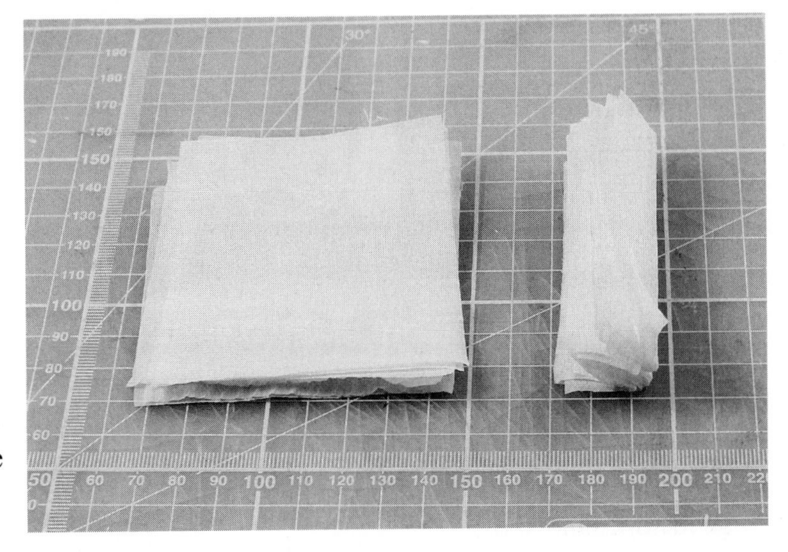

Now that you've made your LED headband, you need to make the paper flowers to complete the look. For my headband, I used white crepe paper, but you can use any type or color of tissue paper you like.

Start by cutting your tissue paper into 3- or 4-inch squares. If you want to make larger flowers for a more dramatic look, you can use bigger sheets. Depending on the paper you're using, you'll need 6 to 10 squares of paper for each flower. My headband has four flowers made of eight sheets each, so I cut out 32 squares.

Stack your squares on top of each other in piles of 6 to 10 squares, and line them up neatly. Next, fold one edge of the stack a quarter of the way into your square, and then turn your tissue paper over for another quarter turn. Fold and turn the paper another two times until you have an accordion-style strip of tissue paper that is one-quarter the thickness of your original square, just as in the picture.

Making Your Paper Flowers

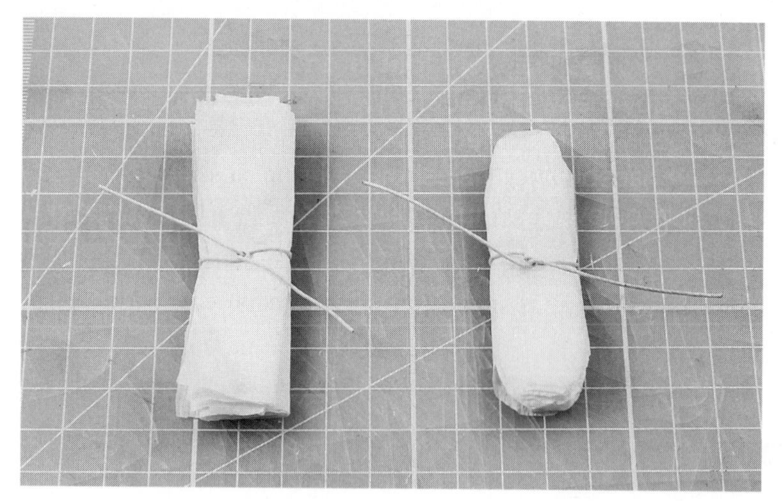

Use your wire cutters to trim off a length of paper- or plastic-covered wire about 5 inches long. Carefully wrap the wire around the center of your folded paper, and then twist it twice to secure it. You should have two loose ends of wire sticking out, which you'll use to attach your finished flower to your head band.

Next, use your scissors to shape the ends of the tissue paper strip into a rounded edge. This will give the tissue paper petals of your flowers a natural, rounded look when they are unfurled. You don't have to worry about your rounded edges looking perfect at this point. Any rough edges on individual petals will blend together and look awesome as a whole when you've finished.

Finishing Your Paper Flowers

Starting on one side of your tissue paper flower, carefully open up the folds. Begin by gently opening up the "accordion" to show your zigzag folds, just like the petals on the right-hand side of the picture.

Next, separate the tissue paper sheets and unfold the individual petals of your flower. To do this, pull gently at the center of each sheet of paper. If you pull from the edges, you may rip your flower. If you do get a few rips here and there, it's not a big deal. When you're done, your paper flower will have enough volume to hide any little mistakes.

Once all your tissue paper sheet petals are unfurled, pull the two paper-covered wire strands together so that they are sticking out at the bottom of your flower.

Finally, give your flower a gentle fluff with your fingers to make the petals look a little messier and more natural.

Attaching Your Paper Flowers

Once you've finished your first paper flower, start the process again to make the rest of your flowers for your LED crown. I made four flowers in total for my crown, but you can make as many as you like. You could even make flowers of different sizes and colors to create your own unique look.

Once you have prepared all your flowers, you can attach them to your headband using the two paper-covered wire strands sticking out of each flower. Be careful not to catch or dislodge your conductive thread paths.

Simply place the base of your flower on the top your headband with one wire leg on each side. Next, twist the two legs together underneath the headband, and then bend the legs back up to the top of the headband and twist again, securing the flower and hiding the end of the wire underneath the bloom. If you have excess wire, trim it off with your wire cutters.

Repeat this process with all your paper flowers.

Finishing Your LED Paper Flower Crown

Finally, add in your battery to check that your circuit still works. With the LEDs on, you can tweak the positions of your flowers until you get the effect you want.

Try on your LED Paper Flower Crown and show off your hard work to your family and friends. This project is perfect for a special occasion such as a party, wedding, or festival. It looks beautiful during the day, but at night you can surprise everyone by adding the battery and lighting up your accessory.

One word of warning: if there is even a hint of rain, you'll need to take off your LED Paper Flower Crown. Electronics don't play well with water, and neither do paper flowers!

Fix It

Not working? Don't worry! Follow these steps to figure out why, and fix it.

1. Check your power.
   - Is your battery the right way around? Flip it over and see what happens.
   - Has it run out of juice? Try another battery.
   - Is your battery connecting into your circuit? Make sure that your battery fits snugly into the holder and that the conductive paths are connecting to the correct side of the battery.
2. Check your components.
   - Are your LEDs working? It's a good habit to check each component before you add it into a circuit.
   - Are your LEDs securely sewn in place? Loose connections mean that your circuit won't work. Tighten your connections, and try again.
   - Are your LEDs sewn the right way around? All the positive bits of the LEDs should be connected to the positive bit of your battery pack, and all the negative bits of the LEDs should be connected to the negative bit of the power supply.

3. Check your wiring.

- Do you have a short circuit? If your positive and negative paths touch, no matter how slightly, your circuit won't work. Tidy up your loose ends, check your knots and thread for fraying, and restitch any crossing paths.

Make More

Paper flowers and LEDs are a match made in heaven. You can adapt this project to make beautiful corsages, garlands, table centerpieces, bouquets, and even statement jewelry. Experiment with different types of paper, different colors of LED, and different sized squares of paper to mix up the effects.

One of the most colorful and delightful makers in London is Rachel Wong, aka KonichiwaKitty. Rachel has made headbands adorned with kitty ears, pompoms, glittery fluff, and even My Little Pony toys! If you love this headband hack, check out her homemade DIY electronics wearables on Twitter or Etsy. She has a great kitty ears kit that teaches you to solder—the perfect way to level up your headband hacking!

PROJECT 14

Glove Hacks: Touchscreen Glove and BFF Gloves

Make your gloves amazing using your DIY electronics skills

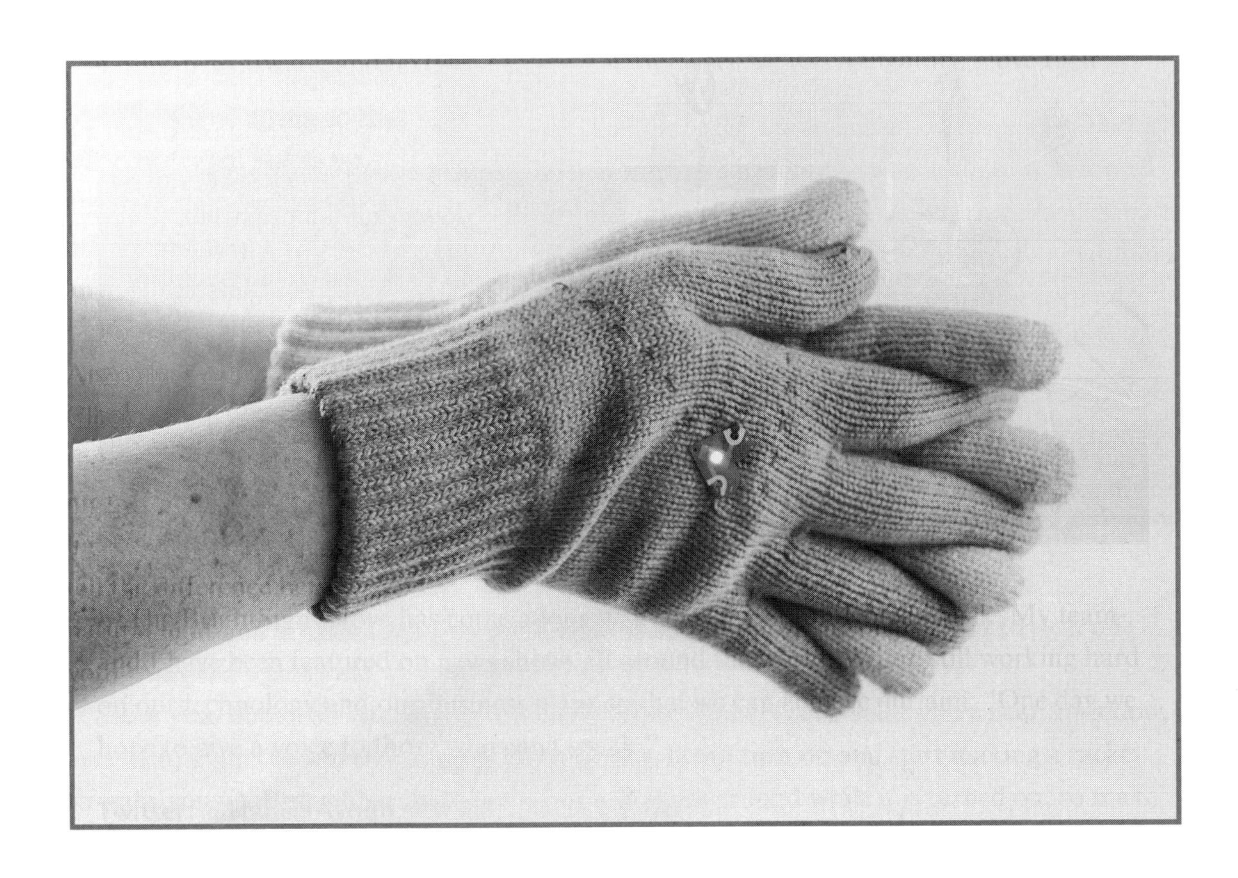

Tools

- One needle
- Sharp scissors

Materials

- Two pairs of woolen gloves
- Two sewable Teknikio heart LEDs
- Two sewable 3V battery packs and batteries
- One square of conductive fabric
- Conductive thread
- Clear nail varnish

This project is two hacks in one. The first hack uses your soft circuit skills to make an ordinary pair of gloves into a smartphone–friendly touchscreen pair of gloves. In the second hack, you will make two pairs of gloves into BFF Gloves that light up when you're holding hands with your best friend.

The BFF Glove was inspired by the work of my BFF, Phoenix Perry. She is a maker and lecturer who loves making and playing games that involve human contact and sensory experiences. You can read more about Phoenix and her work in the Maker Spotlight on page 235.

Preparing Your Materials for the Touchscreen Glove

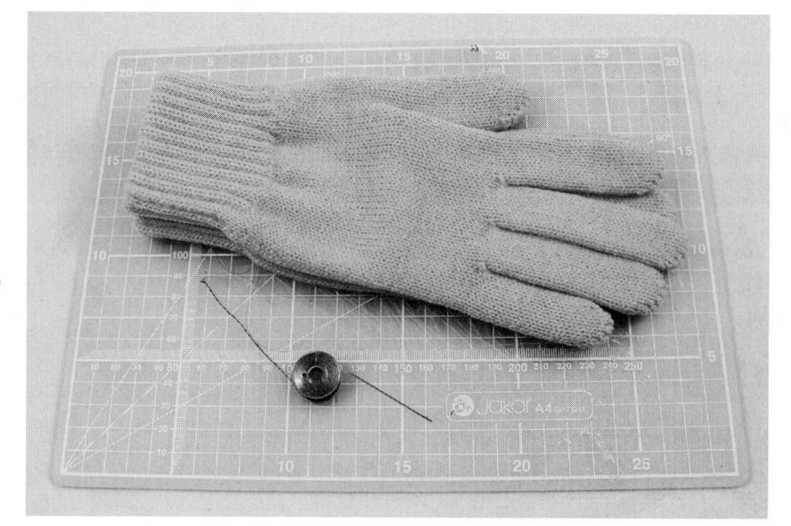

Make sure that you've got all your tools and materials ready. This part of the glove hack only requires a pair of gloves, a needle, and some conductive thread. You may also want to use a thimble to protect your fingertips from getting stabbed while you sew.

For this project, I would recommend that you use a

softer, yarn-like conductive thread instead of the waxier types that you commonly get. Any type of conductive thread will work, but the fuzzy kind will feel nicer on your fingertips.

Before you start sewing, put your glove on, and touch your phone's screen. Notice where the pad of your finger touches the screen, and make a little mark on the fabric of your glove to guide your sewing.

Starting Your DIY Touchscreen Glove

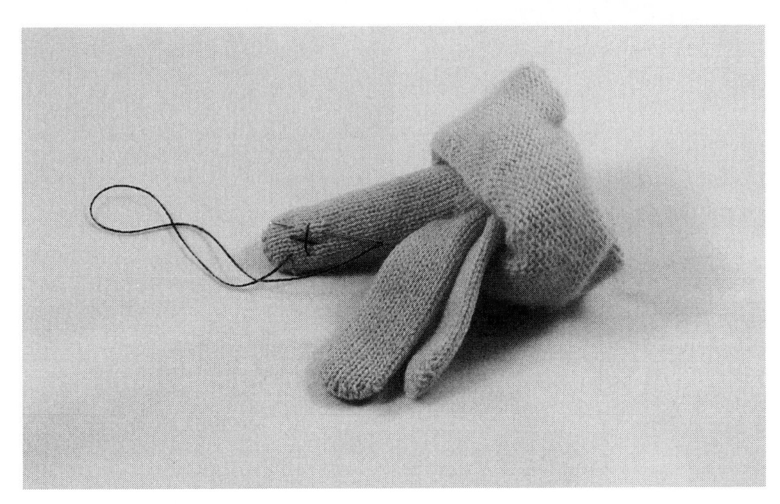

Whatever variety of conductive thread you choose to use, double thread your needle, and tie a knot at the end. If your thread is the slippery kind that tends to come undone, secure your knot with a dab of clear nail varnish.

Carefully roll up the end of your glove so that you can reach the inside of the forefinger with your needle. Push your needle and thread through from the inside to the outside so that the knot is hidden. Start stitching by making an X shape over the little mark on the fabric of the glove that you made earlier.

Rather than pushing your needle in and out of the glove each time, you can keep the needle on the outside of the glove and push it through at a shallow angle, just as in the picture. This is much easier, but you have to be careful to only stitch through one layer of the glove or you won't be able to get your finger in! To help you keep the layers separate, you can sew with one hand and keep a finger from your other hand in the glove. Put a thimble on this finger to stop you from stabbing yourself.

Finishing Your DIY Touchscreen Glove

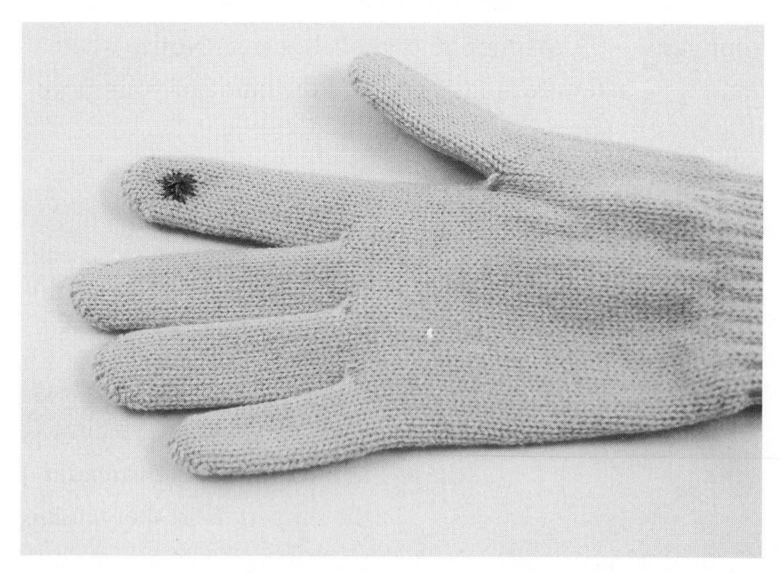

Continue making several stitches until you have a rough star shape, just as in the picture. This star shape will connect your finger to your screen. Again, make sure that your stitches do not go through the other side of the glove. Once you have completed your star, take your needle through to the inside of the glove. Pull the finger inside out, and then tie a knot in your conductive thread, cut the excess thread, and secure it with a dab of nail varnish.

Test your glove on a phone's touchscreen. Can you control your phone through your glove? You may wish to add a conductive thread pad on more than one finger, or you may wish to make the pad bigger or located in a different area to make the perfect pair of DIY touchscreen gloves.

These gloves are a simple hack, but they work really well. DIY touchscreen gloves make a great homemade present in the autumn or winter.

How Touchscreens Work

Smartphones use something called a *capacitive touchscreen*, which can tell when you've touched it. The screen is made of glass that has been coated with a special conductive film. Our bodies are also conductive, so when you press the surface of a touchscreen, you distort the normal electrical field of the phone.

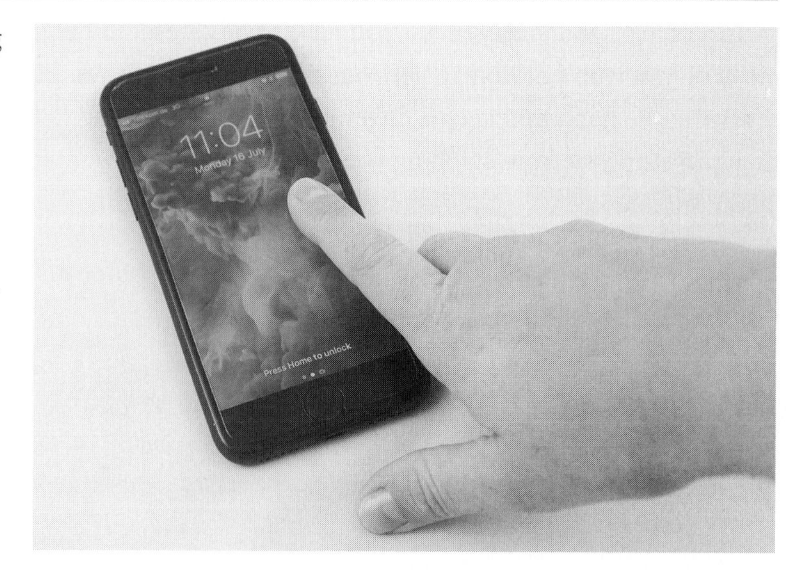

The brain inside the phone recognizes this distortion and uses this information to figure out what you're asking it to do.

You control your phone using the electricity in your fingertips. However, when you're wearing gloves, the nonconductive fabric blocks this electricity from reaching the screen.

As you know, conductive thread allows electricity to flow, meaning that this glove hack allows the static electricity from your fingertips to transfer through to the touchscreen of your phone.

Preparing Your Materials for the BFF Glove

Make sure that you've got all your tools and materials ready. For this part of the glove hack, you'll need two pairs of gloves, two sewable Teknikio heart LEDs, two sewable battery packs, a needle, conductive thread, scissors, and a square of conductive fabric.

This is the first time we've used conductive fabric in a hack. It conducts electricity, just like the conductive thread we've been using. You can use conductive fabric in all sorts of cool projects when you need to make a larger area conductive. You can also use scissors to cut conductive fabric into all sorts of shapes, making it ideal for imaginative DIY electronics projects.

How to Choose and Use Conductive Fabric

There are lots of different types of conductive fabric you can get, and they all behave in different ways. Some of them are made with copper, some are made with silver, some are stretchy, some are sturdy, some are flat, and some are ribbed.

One of the major differences you want to look for when buying your conductive fabric is woven versus nonwoven.

Many fabrics are woven, which means that they are made up of lots of tiny strands. When you cut the edge of a woven material, it will start to fray and come undone. This is why we put hems on our clothes, so that the material doesn't unravel and look messy.

We're been using felt in our projects because it is a very easy material to work with. It is not woven. Just like paper, felt is made by pressing fibers together. This means that you can cut the edge of a piece of felt, and it will not fray.

You can choose to use any conductive fabric in your BFF Gloves. I've chosen to use a nonwoven fabric called *EeonTex* so that I can be sure that my material won't fray. If you use a woven conductive fabric, just make sure you hem it first so that it does not unravel and alter your circuit.

Starting Your BFF Gloves

Take your square of conductive fabric, and cut it into four slim strips, just as in the picture. These strips will make up the switch on your gloves, so you need to figure out where to put them.

Put your gloves on, and ask your BFF to put on the other pair. Now hold hands. Which two hands do you want to connect with: your left and his or her right or the other way around? When you hold hands, do you mirror each other's palms and link fingers or do you put your hands at an angle?

Once you know how you hold hands, you can design the layout of your BFF Glove. When the two gloves are held together, you want the two parallel lines of fabric to cross each other like a *hashtag* (#). This shape is also known as a *pound sign* in the United States.

I tested my glove with my friend, and we held hands with mirrored palms and linked fingers, so I laid out the strips of conductive fabric on my BFF Gloves just as in the picture.

Sewing Your Positive Paths

Now that you know how you're going to lay out your switch, remove the strips of conductive fabric and start sewing your circuits. Thread your needle with conductive thread, and tie a knot at the end.

Put your sewable battery pack on the inside of the wrist on one of your BFF Gloves. Securely sew both positive bits of the battery pack in place, and then take your stitches up and around to the front of the glove.

Rather than pushing your needle in and out of the glove each time, you can keep the needle on the outside of the glove and push it through at a shallow angle. This is much quicker, but be careful to only stitch through one layer of glove or you won't be able to get your hand inside when you're done!

Stitch a path to the middle of the front of your first glove, and then securely sew the positive bit of your sewable LED in place. Take your needle through to the inside of your glove, and then tie a knot in your conductive thread, cut the excess thread, and secure it with a dab of nail varnish. Repeat this process for your other BFF Glove.

Sewing Your Negative Paths

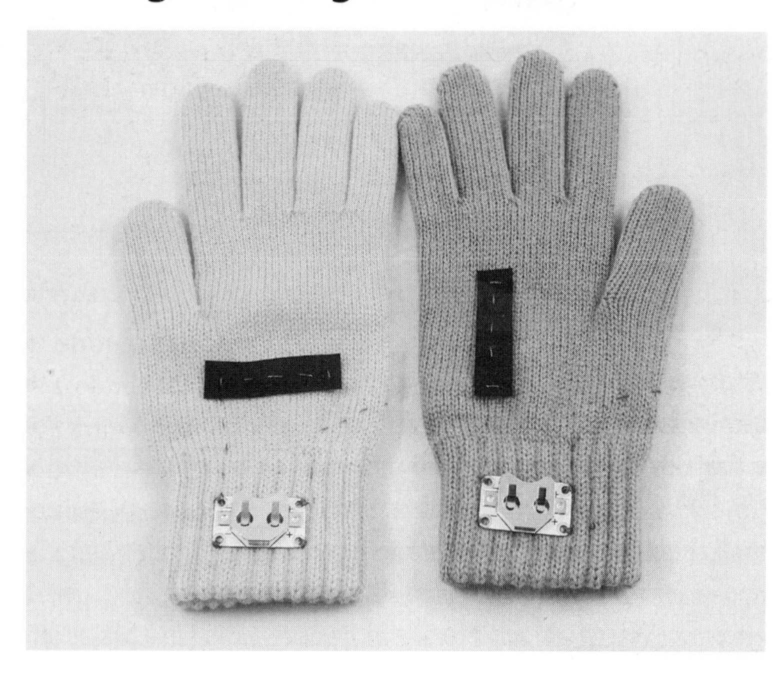

Thread your needle with conductive thread, and tie a knot at the end. Take one of your BFF Gloves and securely sew both negative bits of the battery pack in place. Then take your stitches up past the wrist of the glove until you reach the place where the first strip of conductive fabric should go.

Stitch the strip of conductive fabric to your glove. Depending on the type of fabric you chose to use, this can be a little tough. Use a thimble to help you push your needle through both layers of fabric without hurting your fingers. Again, make sure that you only stitch through one layer of glove.

Once your strip of fabric is securely in place, take your needle through to the inside of your glove, and tie a knot in your thread. Repeat this process for your other BFF Glove. Remember that they will need to form a hashtag shape when you hold hands, so place your conductive fabric strips like you planned at the beginning of the project.

Sewing the Negative Side of Your LED

Thread your needle with another length of conductive thread, and tie a knot at the end. Starting from the inside of your gloves, securely sew the negative bit of your LED onto the front of your glove.

Once your LED is in place, take your stitches away from the negative side of your LED toward one edge of your glove. Depending on how you've planned your BFF Gloves, you will probably take the path from the negative side of the LED in opposite directions on each of the two gloves.

Whichever way your paths go, make sure that you don't get too close to your positive path or your circuit won't work.

Sewing the Negative Side of Your LED

Take your needle to the edge of your glove, and then flip over your gloves to reveal the underside, where the conductive fabric strips are sewn. Take a moment to check that you're going the right way. You want this set of stitches to connect with the strip of conductive fabric that is not yet sewn in place.

Continue sewing until you reach the place where the second strip of conductive fabric should go. Stitch your strip of conductive fabric to your glove. Use a thimble if you are finding it hard to push your needle through both layers of fabric without hurting your fingers. Make sure that you only stitch through one layer of glove.

Once your strip of fabric is securely in place, take your needle through to the inside of your glove, and tie a knot in your thread. Then repeat this process for your other BFF Glove.

Testing and Using Your BFF Gloves

Add your batteries to each glove, and press the strips of conductive fabric to each other so that they form a hashtag/pound sign shape. The LED lights on both hands should turn on because the conductive fabric on each glove is completing the circuit on the other glove.

If it works the first time, you can give one set of gloves to your BFF and keep one for yourself. This hack doesn't just work with holding hands. Can you make the LEDs turn on by high fiving each other?

If your BFF Gloves don't work the first time, see the troubleshooting tips below.

Fix It

Not working? Don't worry! Follow these steps to figure out why, and fix it.

1. Check your power.
   - Are your batteries the right way around? Flip them over and see what happens.
   - Have your batteries run out of juice? Try another battery.
   - Are your batteries connecting into your circuit? Make sure that your batteries fit snugly into the holder and that the conductive thread is connecting to the correct side of the battery.

2. Check your components.
   - Are your LEDs working? It's a good habit to check each component before you add it into a circuit.
   - Are your LEDs securely sewn in place? Loose connections mean that your circuit won't work. Tighten your connections, and try again.
   - Are your LEDs sewn the right way around? All the positive bits of the LEDs should be connected to the positive bit of your battery pack and all the negative bits of the LEDs should be connected to the negative bit of the power supply via the conductive fabric hashtag/pound sign shape.
3. Check your wiring.
   - Do you have a short circuit? If your positive and negative paths touch, no matter how slightly, your circuit won't work. Tidy up your loose ends, check your knots and thread for fraying, and restitch any crossing paths.

Make More

Gloves are a great accessory to hack into a wearable. Your hands are one of the most expressive parts of your body. One of my favorite glove hacks for the winter months is a bike light indicator.

To make a bike light indicator, sew a bright LED to the back of both gloves, connecting one side of your LED to the battery via a sewable button placed on the palm. When you make hand signals to show other road users that you're planning on turning, you can make a fist to touch the button and complete the circuit.

PROJECT 15

Dark-Sensing Amulet

Get +10 courage in dark places with this amulet that lights your way

Tools

- Craft knife
- One needle
- Sharp scissors
- Pencil and paper

Materials

- One 6- × 6-inch square of thick felt in any color
- Conductive thread
- One sewable 3V battery holder with on/off switch
- One 3V battery
- One sewable LED
- One Teknikio sewable light sensor
- Leather or nylon cord

I've always loved adventure and fantasy stories. One of my favorite parts of these kinds of stories is the descriptions of the cool treasure the heroes collect along the way, from Harry Potter's Cloak of Invisibility to Lucy Pevensie's cordial and dagger. These special items help our heroes vanquish enemies, heal friends, and find their way out of tricky situations.

This project is inspired by these fantastical treasures. We'll be using a new type of component called a *light sensor*, which will control our circuit based on how light or dark it is.

Preparing Your Materials

Make sure that you've got all your tools and materials ready. Then prepare your felt. Turn to page 254 for a template of the Dark-Sensing Amulet. Trace and cut out the teardrop shape, and then pin it to your felt and use it as a guideline to mark out your pattern.

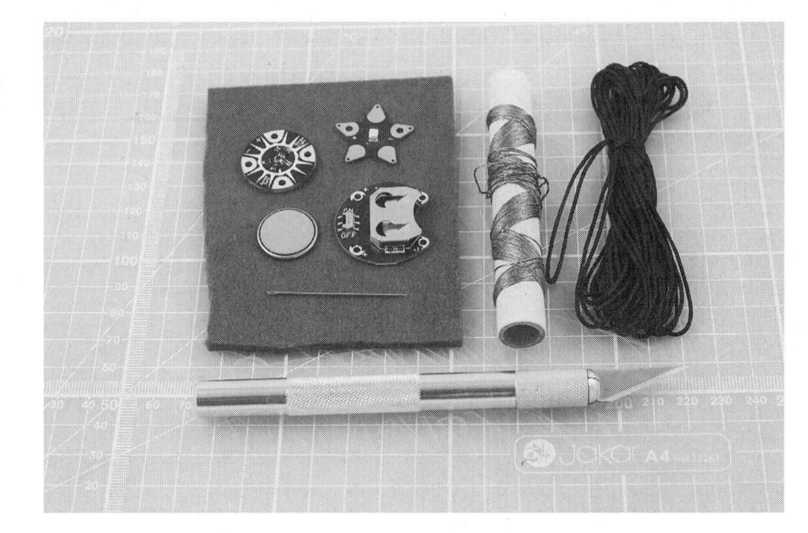

For this project, I used a thick felt because I wanted

the shape to stay in place. If you only have thin felt, you can cut two or three teardrop shapes and sew them together in layers to make a stiffer fabric.

Once you have your shape cut out, use a craft knife to make a small incision at the top of the teardrop, and then use scissors to carefully cut a small hole.

Take your leather or nylon cord, and cut off a piece about 30 inches long. This cord is going to thread through the small hole you just made so that you can hang your amulet around your neck using a special knot called a *lark's head knot*.

How to Make a Lark's Head Knot

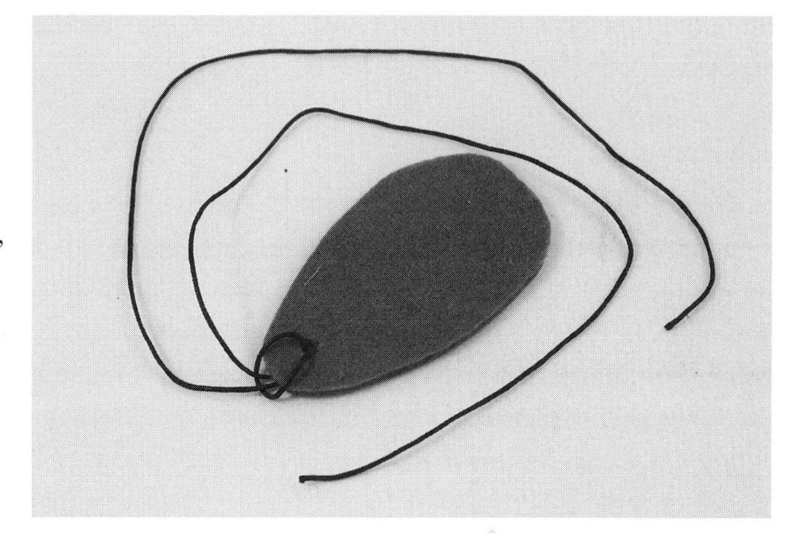

A lark's head knot is also called a *cow hitch knot*. This kind of knot is used to attach a cord to something. It is a simple knot that does not untie or slip. As a bonus, it looks very attractive because it is symmetrical and lets the thing it's attached to lie flat. It is very commonly used in jewelry making.

Fold the cord in half down the middle. You should have two cord ends, with one end looped and one end open. Push the folded end about 1 inch through the small hole you made at the top of the teardrop, and then pull both strands of the open end through the loop.

Pull the whole thing taut, and you're done! Have a look at both sides of the amulet. You'll notice that the knot looks different from each side. There is no "right way round" for this knot, so choose the side of the lark's head knot that you like the look of most to be the one at the front. Leave the open ends as they are for now.

Sewing Your Battery Pack

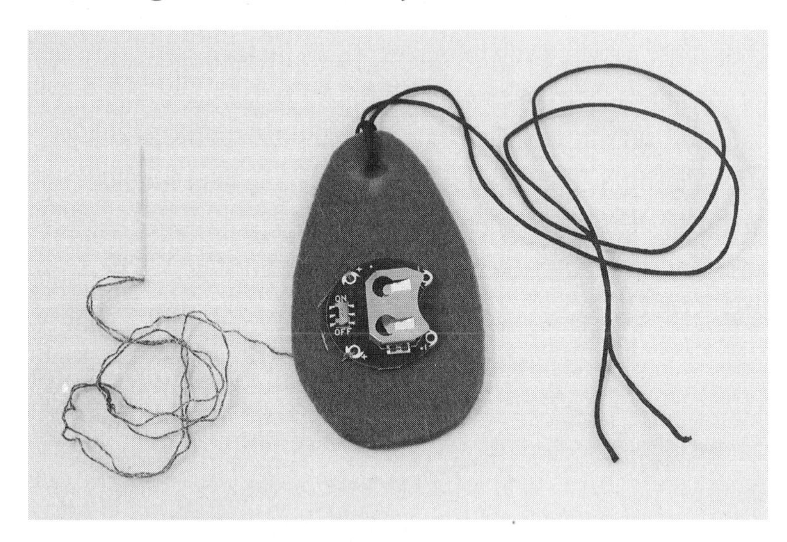

Turn your amulet onto the side you want to be the back, and then place your sewable 3V battery holder in the center of the shape. I've used a Lilypad sewable battery holder with on/off switch for my version of this project.

Thread a needle with a length of conductive thread, and tie a knot at the end. If your thread is the slippery kind that tends to come undone, secure the knot with a dab of clear nail varnish.

Sew the positive side of your battery pack first, stitching the top hole first and then the bottom hole. Because you're sewing with a thicker fabric, you may find it helpful to put a thimble on your thumb to help you push the needle through the fabric.

As always, take care that your thread doesn't get looped or knotted underneath your fabric. Once your battery pack is securely stitched in place, you can start sewing your light sensor onto the front of your amulet.

What Is a Light Sensor?

A *light sensor* detects light levels. Ordinary light sensors (sometimes called LDRs) are small, simple to use, cheap, and hard wearing, which is why you'll find them in all sorts of places, from household appliances to children's toys.

They work by letting more or less electricity through the circuit based on how much light is shining on the face. The face is the bit with the squiggly lines that you can see on the right in the picture.

To make your first project with a sensor as simple as it can be, we're using a special sewable light sensor, which you can see on the left in the picture. This light sensor has two modes, "light" and "dark."

In light mode, the sensor triggers an output to turn on when it senses light and turn off when it senses dark. In dark mode, the sensor will trigger an output to turn on when it senses dark and turn off when it senses light. In our project, we'll be using the dark mode to turn on an LED.

Sewing Your Light Sensor

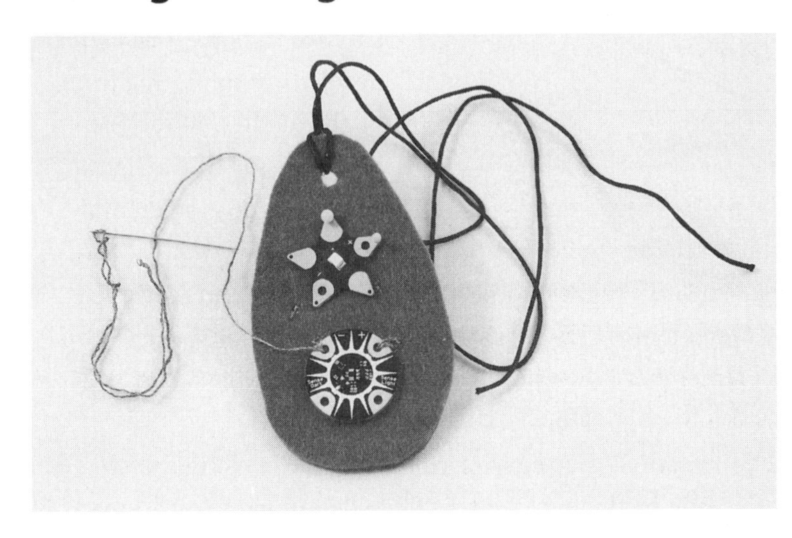

Once the positive side of your battery pack is securely stitched in place, push your needle through to the front of the amulet. The light sensor has four holes: positive, negative, dark, and light. You will be sewing paths to positive, negative, and dark, so you need to make sure that your sensor is oriented in a way that makes sense. I've put my sensor at the bottom of the amulet, with the negative hole at the top left, the positive hole at the top right, the dark hole at the bottom left, and the light hole at the bottom right.

Use your needle to stitch conductive thread from the positive side of the battery pack to the positive side of the light sensor. Then tie off, trim your thread, and secure the knot.

You may find it a little tricky to stitch through the thick fabric, or you might find that the battery pack gets in the way of stitching your sensor. If so, remember that you don't have to go all the way through the fabric. You can angle your needle slightly and make shallower stitches. This technique can be tricky at first but is super useful once you get the hang of it!

Sewing Your Light Sensor to Your LED

Thread a needle with a length of conductive thread, and tie a knot at the end. Flip your amulet back over, and start sewing the negative side of your battery pack, stitching the top hole first and then the bottom hole.

Once the negative side of your battery pack is securely stitched in place, push your needle through to the front of the amulet. Stitch the negative side of the light sensor to the amulet, and then take your path up toward the space for your LED.

I'm using a Teknikio LED star because I think it looks pretty, but if you don't have one, you can use any sewable LED or even an ordinary LED that you make sewable using a pair of needle-nose pliers. If you can't remember how to do that, look back to page 52 for a reminder.

Whichever type of LED you choose, stitch the negative side of the LED using the conductive thread from the negative side of the light sensor. Then tie off, trim your thread, and secure your knot.

Finishing Your Circuit

To finish your circuit, you'll need to connect the positive side of your LED to the "Sense Dark" hole of the light sensor. When making the stitches down to the bottom of the amulet and across to the dark hole, you need to be very careful not to cross any of your paths.

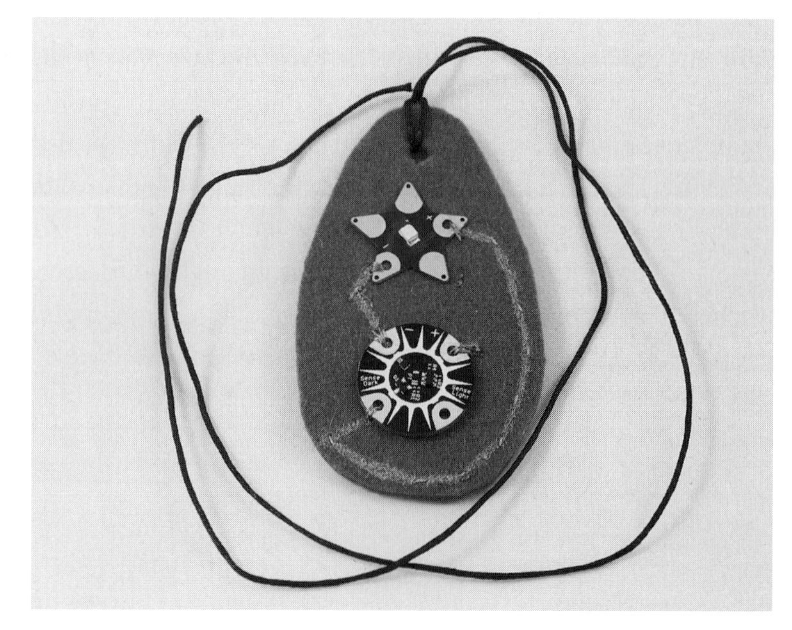

In the picture, I've drawn the path on which I've taken my stitches. They go along the very edge of the amulet so that I don't accidentally touch the stitches next to the battery pack. Take extra care that you don't get any loops or stray threads underneath your amulet.

When you get to the "Sense Dark" hole of the light sensor, sew it on securely, and then tie off, trim your thread, and secure your knot.

This project has a couple of paths quite close to each other. If they touch, your circuit won't work properly. Some types of conductive thread tend to fray, so my tip is to spray your thread with a little hairspray after you've finished to keep the thread from getting frizzy and causing problems.

Now you're ready to make the finishing touches to your Dark-Sensing Amulet.

How to Tie an Adjustable Sliding Knot

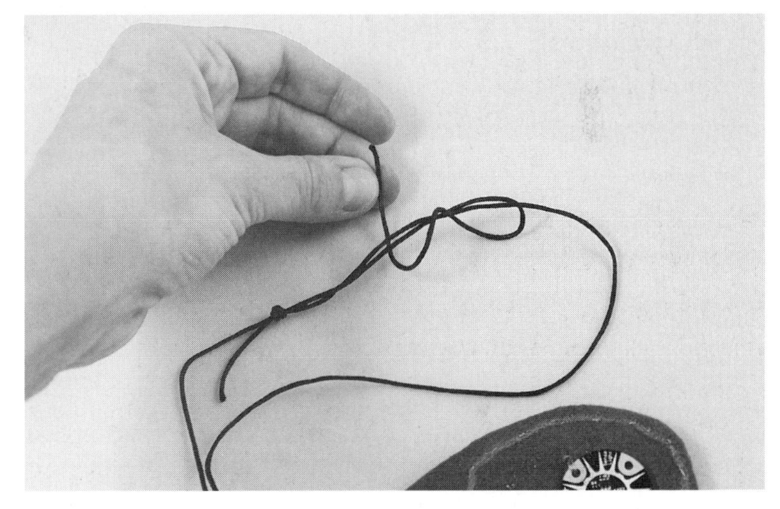

Sliding knots allow you to take your amulet off and put it on easily, and you can also use them to adjust the length of the amulet. To make a sliding knot, you need to tie your cord ends to the cord on the opposite side. Lay your amulet in front of you with one cord end on the left and the other on the right.

Take the end of the cord on the left about 3 inches from the end, and put it over the cord on the right about 3 inches from the end. Bend the left-hand cord over the right cord so that it forms a loop around the right cord. Pinch that loop to keep it in place while you make the knot.

Take the loose end of the left-hand knot and wrap it round both the other bits of cord three times so that you have a little coil just as in the picture. Finally, poke the end of the left-hand pink cord through those loops and back out the right end. Pull this tight, and you should have your first sliding knot.

Repeat this process on the other side, and you should have two sliding knots that you can adjust as you like.

Using Your Dark-Sensing Amulet

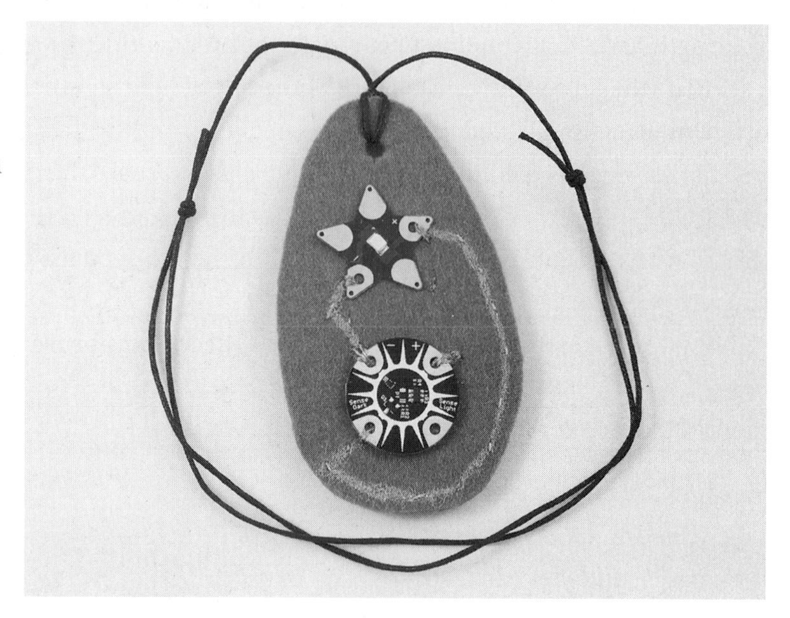

You have now finished your Dark-Sensing Amulet. Put your 3V battery into your battery holder, flip the switch to "on," and put the amulet around your neck.

Next time you go somewhere dark, it should light up to show you the way. I tested my amulet in my wardrobe, but sadly I didn't make it to Narnia.

You can also test it by holding your finger over the light sensor. There are a few different tiny parts mounted on top of the sensor. Can you figure out which bit you need to cover to trigger the sensor?

Fix It

Not working? Don't worry! Follow these steps to figure out why, and fix it.

1. Check your power.
   - Is your battery the right way around? Flip it over and see what happens.
   - Has it run out of juice? Try another battery.
   - Is your battery connecting into your circuit? Make sure that your battery fits snugly into the holder and that the conductive thread paths are connecting to the correct sides of the battery.
2. Check your components.
   - Is your LED working? It's a good habit to check each component before you add it into a circuit.
   - Is your LED securely sewn in place? Loose connections mean that your circuit won't work. Tighten your connections, and try again.
   - Are your components sewn the right way around? The negative bit of the LED should be connected to the negative bit of your light sensor, which should be connected to the negative side of your battery pack. The positive bit of the LED should be connected to the "Sense Dark" bit of your light sensor. The positive bit of the light sensor should be connected to the positive bit of your battery pack.

3. Check your wiring.

- Do you have a short circuit? If your paths touch, no matter how slightly, your circuit won't work. Tidy up your loose ends, check your knots and thread for fraying, and restitch any crossing paths. Use my hairspray trick to keep close paths from fraying and touching each other.

Make More

If you live in a part of the world with dark winters, you can attach this amulet onto your bag or coat to help you be safe and seen on your way home from school. You can also unpick your light sensor from the amulet and reuse it in loads of different ways. In this project, we used the "Sense Dark" function. What project can you think of that would use the "Sense Light" function? Check out the robotic alarm clock in Project 19 for one way of reusing this useful component.

PROJECT 16

Secret Signal Mood Badge

Mix up your colors and secretly signal your emotions to your friends

Tools

- Craft knife
- One needle
- Sharp scissors
- Pencil and paper

Materials

- Two 6- × 6-inch squares of felt in any color
- Conductive thread and ordinary thread
- One 3V battery
- One sewable RGB LED from Teknikio
- Three sewable on/off switches
- One sewable badge pin

Do you ever wish that you could tell your family or friends how you feel without saying anything? With the Secret Signal Mood Badge you can tell someone that you're hungry, grumpy, full of joy, tired, mischievous, or bored with the flick of a switch.

In this project, we're using a new type of LED, one that can show red, green, blue or a combination of any of those three colors. We're combining this RGB LED (short for red, green, and blue light-emitting diode) with three on/off switches to play with color combinations.

Some RGB LEDs need different wiring, so be sure to read the information about RGB LEDs on the next page carefully before sewing your project.

Preparing Your Materials

Make sure that you've got all your tools and materials ready. Then prepare your felt. Turn to page 255 for a template of Secret Signal Mood Badge. Trace and cut out the two diamond shapes, the two tabbed circle shapes, and the circle shape.

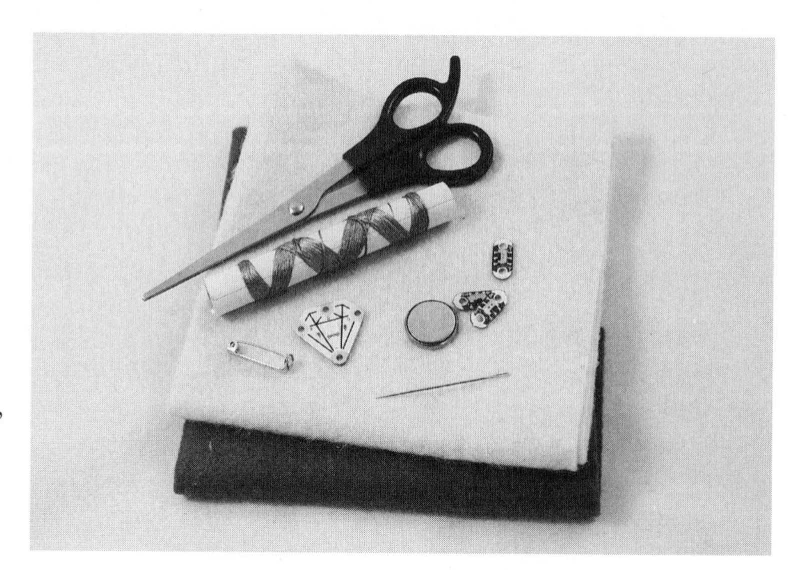

Pin each part of the template to your felt, and use it as a guideline to mark out and cut your pattern. I chose to use two contrasting colors in my badge, but you can choose whatever color of felt you like.

Once you've cut out your shapes, place the smaller diamond on top of the larger diamond, and then use a needle and ordinary thread to sew the two diamonds together. This double layer of felt looks cool, but it also adds strength to your badge.

Now that you've got the base of your badge ready, you can move on to sewing your RGB LED.

What Are RGB LEDs?

So far we've been working with LEDs that have only one color. An RGB LED is actually three tiny LEDs—one colored red, one green, and one blue—in one neat package.

Our eyes have three types of light receptors in them: red, green, and blue. By changing the combinations of red, green, and blue light, you can make any color visible to humans. This is because our brain processes the RGB light and converts it into a color.

Normally, RGB lights are controlled by a chip or a computer. We're keeping it simple and controlling our RGB LED with three on/off switches instead of code. This means that we can't make every single color in the spectrum, but we can create eight color combinations for our Secret Signal Mood Badge.

In the pictures for this project, I'm using a sewable Teknikio RGB LED, but you can also use a sewable Lilypad RGB LED. If you do use a different RGB LED, pay attention to the positives and negatives. The Teknikio RGB LEDs have a shared negative hole, also called a *common cathode*. However, the Lilypad RGB LEDs have a shared positive hole, also called a *common anode*. My wiring instructions are for RGB LEDs with a shared negative hole—if you have a different type, just turn the battery around. Easy!

Wiring Your RGB LEDs

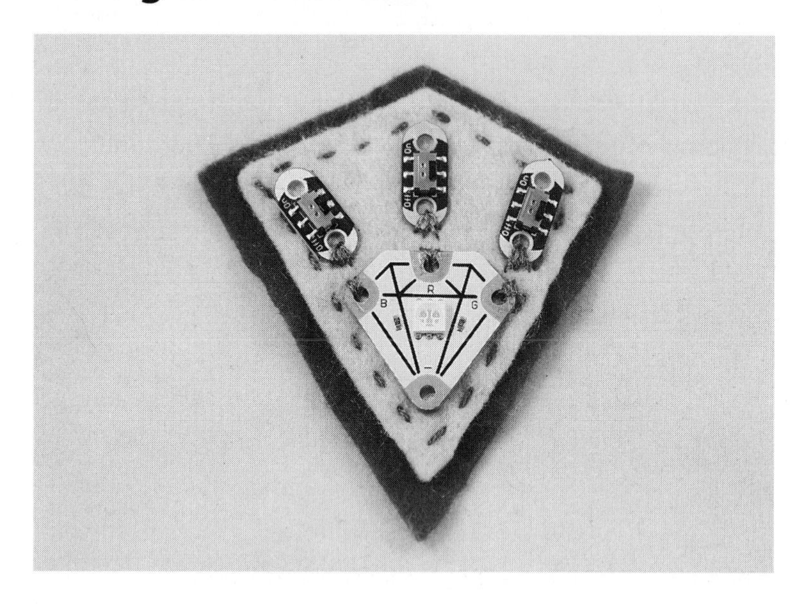

Thread a needle with a length of conductive thread, and tie a knot at the end. If your thread is the slippery kind that tends to come undone, secure your knot with a dab of clear nail varnish.

We're going to secure our RGB LED and three on/off switches in place with three sets of separate stitches. Sew the B bit of your RGB LED first, and then take your path to the "off" bit of your first switch. Secure this tab in place, and then tie off the thread, trim the thread, and secure your knot. Because the Secret Signal Mood Badge is quite small, you will have to pay attention to making your stitches neat and tidy, with no knots or loops underneath your fabric.

Next, repeat this process with new conductive thread, attaching the R bit of the RGB LED to the "off" bit of the central on/off switch. When this is in place, tie off the thread, trim the thread, and secure your knot.

Finally, do it all again with the G bit of the RGB LED to the "off" bit of the last on/off switch.

Finishing Your Positive Paths

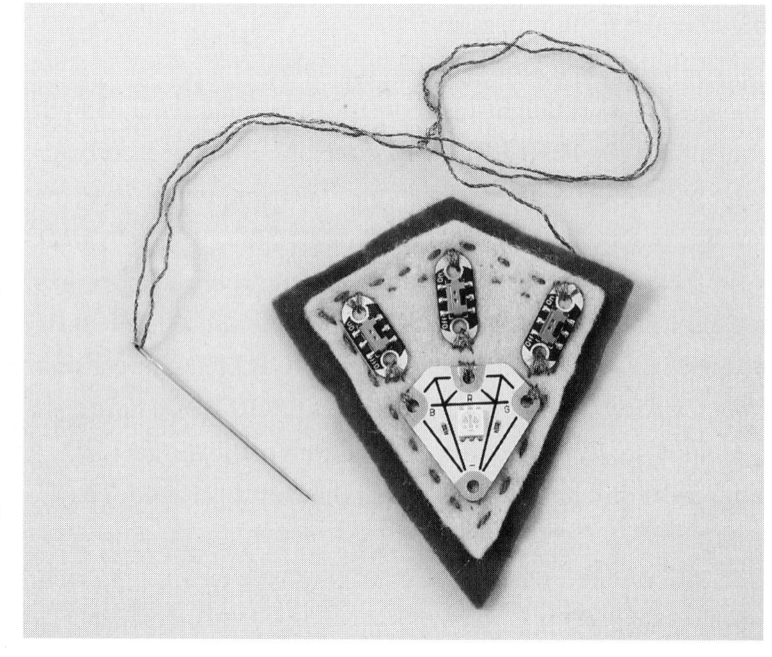

The next thing we need to do is connect all three of the on/off switches to each other at the "on" end and then connect them all to our DIY battery pack.

Thread your needle with another length of conductive thread, and tie a knot at the end. Start sewing the "on" bit of the on/off switch on the left-hand side. Once that is securely in place, take your stitches over to the central on/off switch and sew the "on" bit in place. Next, stitch your way to the final on/off switch on the right, and sew the "on" bit in place.

Finally, stitch back to near the center at the top, and push your needle through to the back of the diamond. You're now ready to sew the positive side of your DIY battery pack.

Sewing Your Battery Pack

Push your needle through a tab on one of the fabric battery packs you cut out earlier. Secure the tab in place with a couple of stitches. Now use a running stitch to sew a path to the center of the tabbed circle. This running stitch should not go through the fabric making your diamond. Your stitches should just be on the tabbed circle. This is because we'll be putting another bit of fabric underneath the tabbed circle to stop our paths from crossing.

Next, make several stitches in a rough star shape, just as in the picture. This star shape will be where one side of your battery connects. Again, your stitches should not go through the fabric making your diamond—they should just be on the tabbed circle.

Finally, tie off and trim your thread, and then secure the knot.

How to Avoid Short Circuits

Throughout this book, I've been telling you to be careful not to cross paths or let fraying ends touch each other. If they do, your circuits will not work, as you've probably already experienced!

The reason crossed paths and fraying threads mess up your projects is that they create short circuits. Electricity will always take the easiest path, so if you create a shortcut—a short circuit—your electricity won't go where you want it to. A short circuit that doesn't go through any components can lead to your battery overheating dangerously!

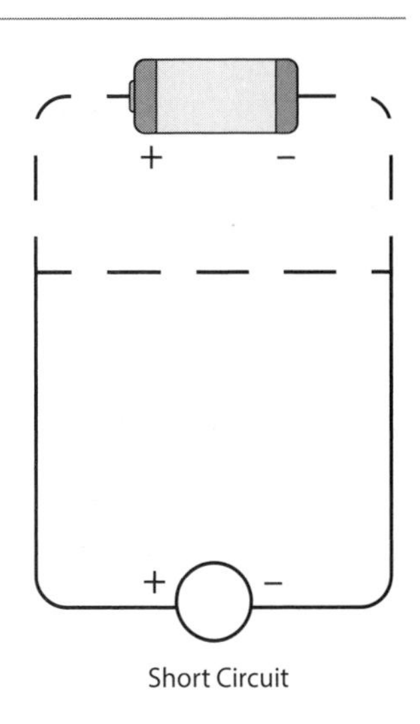

Short Circuit

To avoid short circuits, we can do two things. First, we can be careful about keeping our paths away from each other so that they don't cross or accidentally touch. The other thing we can do is insulate our paths so that they cannot touch each other.

We actually used insulation back in Part One, in the Pop-Up Cityscape project (Project 6). We used a nonconductive material—card—to allow us to cross two paths. Normal wires are insulated with a coating of colored plastic, with the conductive metal underneath.

Insulating Your Circuit

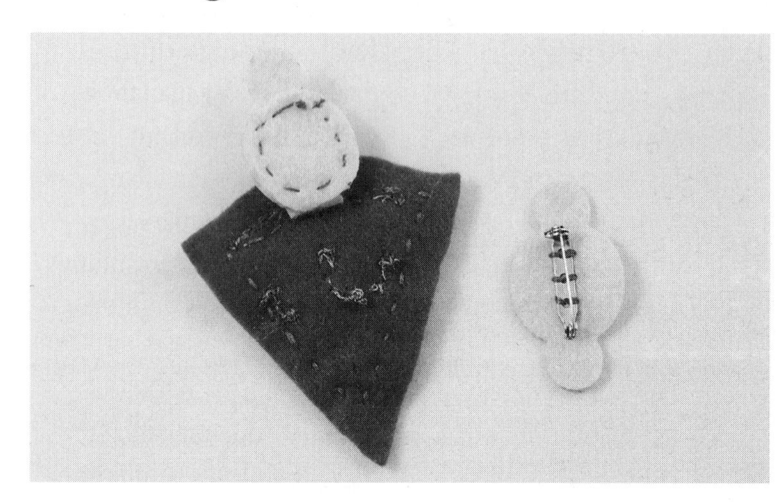

In this project, we'll be using an extra layer of felt—a nonconductive material—to insulate the conductive thread paths of our circuit. To do this, you simply need to thread a needle with ordinary cotton thread, and then sew the circle onto the underside of the battery pack tabbed circle, just as in the picture. Once you've sewn all the way round the circle, tie off your thread, and trim the end. This is all there is to insulating your battery from the rest of the circuit! Take a look to make sure that no stray bits of conductive thread are peeking out on the underside of the battery pack.

Finally, use another length of ordinary cotton thread to attach your sewable badge pin to the other tabbed circle. This tabbed circle will make up the other side of your battery pouch.

Finishing Your Secret Signal Mood Badge

Place the second tabbed circle shape on top of the first. Push the needle connected to the negative hole of your RGB LED through both layers of tabs on the opposite side to the first tab you sewed. Secure the tab in place with a few stitches, and then sew a path to the center of the tabbed circle. Make sure that you only sew on the top layer, avoiding the bottom tabbed circle that you've already sewed. When you reach the center of the tabbed circle, make several stitches in a rough star shape. Tie off your thread.

Thread a needle with ordinary cotton thread. Place a 3V battery between the two layers of tabbed circles so that you know the size. Then use your cotton thread to sew one tab first, then in a semicircle around the bottom of the pouch to the other tab. Sew the second tab in place, and test the size of the battery holder with your 3V battery. If the pouch is too loose, sew a little up the sides until they are snug but still removable. When you're happy with your DIY battery pouch, tie off your thread.

Using Your Secret Signal Mood Badge

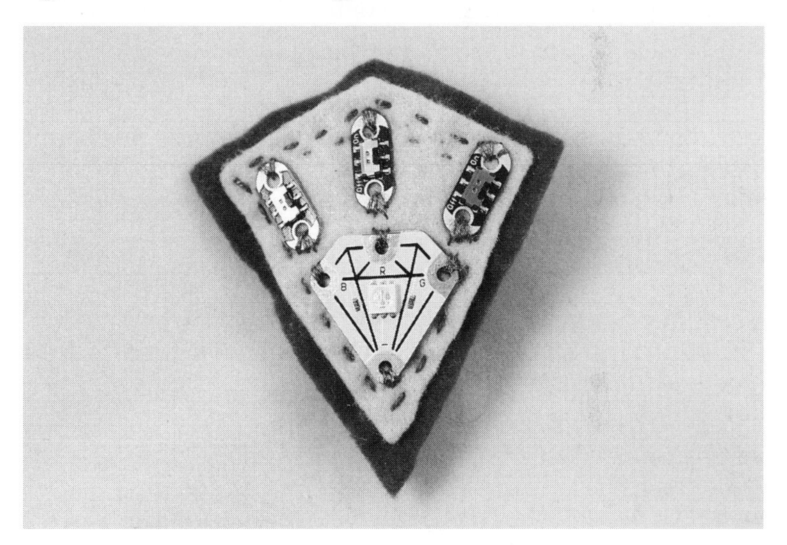

Make sure that all your on/off switches are flipped to "off." Put your battery in the DIY battery pouch, and flip "on" the switch connected to R. Your light should turn red. If it doesn't, check the troubleshooting guide below to help you find your problem and fix it. Next, flip off R and flip on G. Your light should turn green. Finally, flip off G and turn on B. Your light should turn blue. You now have three colors, but you can get your badge to show other colors by using different on and off combinations.

We can figure out how many possible colors you have using something called *binary*. The first switch has two options: on and off. In binary, this is shown as 0 (off) and 1 (on). Now let's add the second switch. Our possible combinations using both switches are on + on (1,1), off + off (0,0), on + off (1,0), and off + on (0,1). This is double the amount of options! Use the above-mentioned 0 and 1 system to add in the third switch and find out how many color combinations your Secret Signal Mood Badge has. These zeros and ones are the building blocks of all code and all computers. Cool, eh?

To use your Secret Signal Mood Badge, assign different moods or emotions to each of the colors. Then share the key to your secret code—if you want to!

Fix It

Not working? Don't worry! Follow these steps to figure out why, and fix it.

1. Check your power.
   - Is your battery the right way around? Flip it over and see what happens.
   - Has it run out of juice? Try another battery.
   - Is your battery connecting into your circuit? Make sure that your battery fits snugly into the holder and that the conductive thread is connecting to the correct side of the battery.
2. Check your components.
   - Is your RGB LED working? It's a good habit to check each component before you add it into a circuit.
   - Is your RGB LED securely sewn in place? Loose connections mean that your circuit won't work. Tighten your connections, making sure the R, G, and B paths are not touching each other, and try again.
   - Is your RGB LED connected the right way around? Check the labels on your component because it might be different from my Teknikio RGB LED.
3. Check your wiring.
   - Do you have a short circuit? If your positive and negative paths touch, no matter how slightly, your circuit won't work. Tidy up your loose ends, check your knots and thread for fraying, and restitch any crossing paths. Check that the insulation pad at the back of your battery pouch is doing its job. You can also try my hairspray trick from Project 15 to keep nearby paths from fraying and touching each other.

Make More

If you enjoyed making color combinations with your RGB LED, you could take it one step further by using code to control your colors. Try out Circuit Playground Express by Adafruit, which has RGB LEDs and loads of other cool components baked in. You can code it using simple blocks, so it's great for beginners. A good book to to help you on your way is Getting Started with the Adafruit Circuit Playground Express *by Mike Barela.*

Maker Spotlight

Name: Hadeel Ayoub
Location: London
Home: Saudi Arabi
Job: PhD researcher at Goldmsiths, University of London

Hadeel Ayoub is the founder of BrightSign, a wearable technology startup developing enabling technology dedicated to helping speech-disabled children with nonverbal autism and their parents. Her glove translates sign-language gestures into speech in real time. She has won many global awards in the category of technology innovations for social care and community impact. Hadeel is working with schools to implement her technology in classrooms to enable the inclusion of deaf and autistic children.

Q. ***What do you like to make and craft?***
A. Innovative tech for sure! I love new ideas and new applications. I'm specifically interested in gesture recognition and human-computer interaction. My very first attempt that led to BrightSign was to interact with a computer without the classic mouse and keyboard.

Q. ***What are you currently working on?***
A. I'm still working on BrightSign but with different applications. I am planning a new glove that is dedicated to teaching toddlers how to sign. The screen will display images, and the speech will be in their moms' voices. We are also making new fun patterns for the glove that include kids' favorite characters and themes with interactive LED lights for feedback.

Q. ***What are you doing next?***
A. I am continuing to get business training to support my technical skills. I want to grow my team and retain the great people who are currently working with me to make the gloves the best they can be.

Q. ***What is your dream project?***
A. I dream about designing a technology that can be an extension of the senses. I really do want to follow my dream of "one day giving a voice to those who cannot speak." I want the users to have full control over how it is done.

Q. ***Who inspires you?***

A. My mother in her determination and commitment to overcoming challenges: educational, social, and economical. She decided to study medicine in an era and a country where girls were not allowed to be educated. She left the country to follow her dream and came back as a doctor only to face social stigma. She married a foreigner and was outcast from society, and as a result, she lost all government benefits. She is now a pioneer in her field and made a family of doctors and innovators.

Q. ***Can you share your favorite books, websites, or places you go to learn new skills?***

A. I am interested in history, art, and architecture, so I tend to read a lot of art history books. My favorite author is Ken Follett, and my favorite book is one book of his epic history trilogy *Pillars of the Earth*, which documents that the building of the first-ever chapel took 107 years and multiple generations. To learn new skills, I visit DIY tech platforms online. I also go to a lot of tech meet-ups to see what other people are doing. I also draw inspiration from art. I enjoy mixing art and technology together. To exhibit my tech gloves at an art exhibition, I learned how to cast hand models and cast my own hand on which to display the gloves.

Q. ***What is your favorite place to get materials or tools?***

A. I do most of my electronics shopping online. Cool Components, The Pi Hut, and Pimoroni are my favorites. In addition to buying electronics for work, I buy electronics to play with on weekends with my kids.

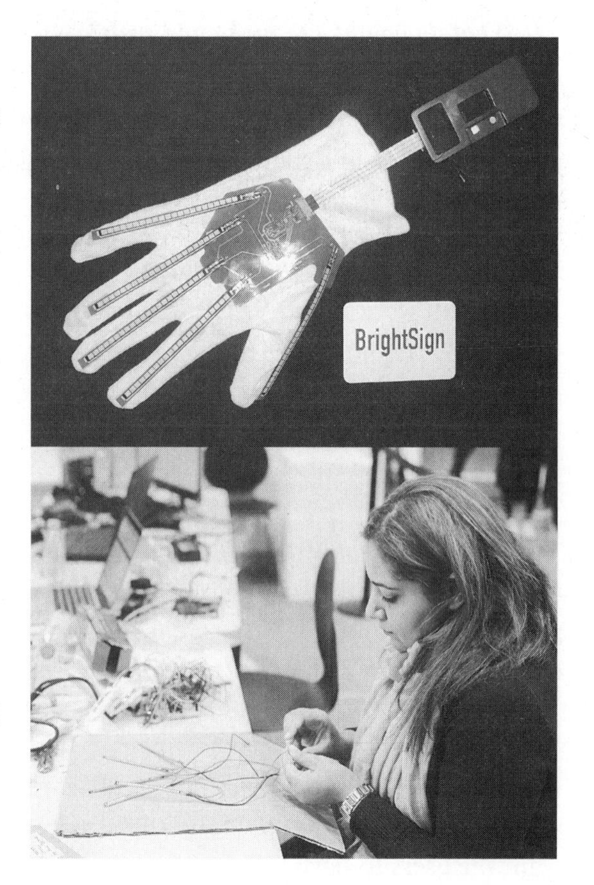

BrightSign Glove

BrightSign is a technology startup that develops assistive technology solutions to enable communication for individuals with nonverbal disabilities. Our first product is the BrightSign Glove, a smart glove equipped with multiple sensors and machine-learning software that translate hand gestures to text and speech. It is designed to enable individuals who use sign language as their primary language to communicate directly with the public without the need of an accompanying translator.

The BrightSign Glove started life as a smart glove that let you draw in the air and then see your sketch on a computer. I competed in a "hackathon" in South Korea and changed the output from sketches to sign language. My team won the hackathon and was in the news all around the world. My inbox was flooded with emails from parents, teachers, and therapists who wanted to buy the glove, so I decided to develop the idea further.

I have made 11 working prototypes to date, all designed based on the feedback of users. I started with an Arduino Uno, flex sensors, and a four-digit numerical screen. Prototyping has been a long process, but it is very rewarding. I was blown away when the glove said its first words! Once I had an iteration that included my developments in hardware, software, and textiles, I tested it with 50 people who taught the glove their own signs.

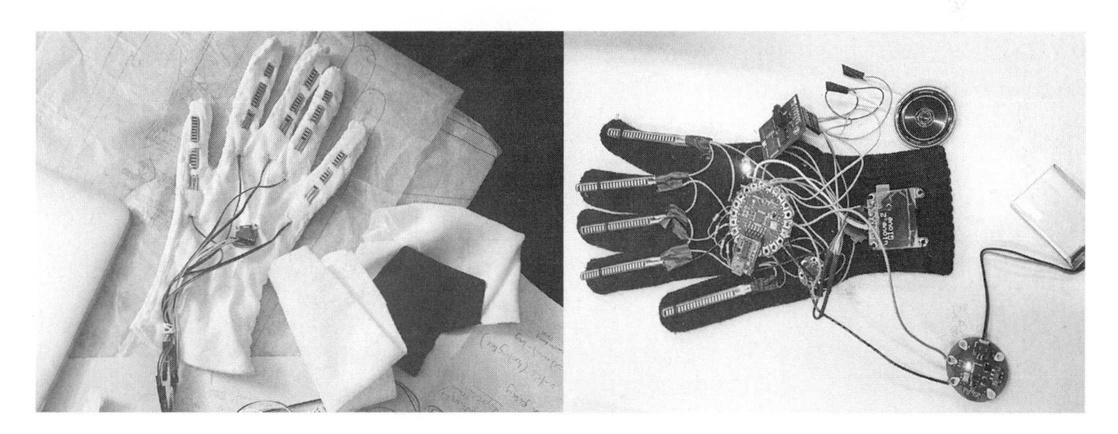

The BrightSign Glove has come a long way since my South Korean hack. My team and I have been featured on news shows all around the world. We are still working hard on our technology and our business plans so that we can achieve our aim: "One day we hope to give a voice to those who can't speak."

Twitter: @HadeelAyoub
Website: https://brightsignglove.com

PART FOUR

Robots

Introduction to Robots

Welcome to the Robots part. In this part, we are going to explore what makes a robot a robot and then make some simple robots using DIY electronics and paper engineering.

Robots are machines that do a set of things automatically, meaning that they act without humans directly controlling them. Robots usually have a set of inputs, a set of outputs, and a "brain." An *input* is thing that triggers something, such as a sensor or a switch. An *output* is something that does something, such as lights, sound, or movement. The "brain" of a robot is designed to act, sense, or react. Most robots these days are controlled by computer program "brains" on microcontrollers or computer chips, but you can also make robots react to sensors and trigger outputs using circuitry and electronics.

The world of robotics is changing fast. You can find robots in every part of the world doing almost any kind of job you can think of. Robots work in factories building, delivering, and checking products. Robots explore places hostile to humans, such as the deep sea or sites with dangerous nuclear waste. Robots even look after elderly people and children. Do you think there are jobs a robot can't do? What about creative jobs? In this part, we're going to make a Modern Art Robot to test our assumptions about creativity and automation. What do you think now? Can a robot make art?

Because this is probably the first time you've made a robot, I'm going to be keeping our first robotic projects very simple. We are not going to be using any code or logic, but if you want to go further and make robots with code and electronics, here are lots of great beginner microcontroller "brains" made especially for robotics. Check out Crafty Robot's SmartiBot, Adafruit's Crickit, or the BBC Micro:bit for the next stage of your robotics journey.

Robots can do all sorts of things, but over the course of this part, we're going to concentrate on one of the most exciting outputs in robotics—movement! We're going to use motors for the first time, figure out how to change the movement of our robots with design, and then make a paper engineering automaton. Here's where to find this information:

Robots: Essential Skills

We will also learn how to use a new DIY electronics tool in this section and find out about making circuits with wires instead of copper tape or conductive thread. Here's where to look up this information:

Robots: Essential Skills

What Are Motors?

There are lots of different inputs (things that trigger something) and outputs (things that do something) that you can use in robotics, but we're focusing on motors and movement.

We use motors for movement, from making robot arms wave to driving cars. We're going to be using a type of motor called a *direct-current (DC) motor*. You can't control DC motors really precisely, but you can get them to spin really fast, making them good for movement. We'll be hacking some little 3V DC motors into vibration motors to get our simple robots to move. A *vibration motor* is a type of DC motor with an unbalanced weight on one end.

Motors can require a lot of power. Some need to be wired up in a special way with other components, so you do need to be careful about which ones you buy. The little motors we are going to use are powerful enough to make our robots jump around but not powerful enough to move anything heavy. This is because our motors can spin really fast, but they don't have a lot of *torque*, or *turning power*.

How to Use Wire for the First Time

So far in this book we have used copper tape and conductive thread to make our circuits. However, sometimes the best option for making a circuit is wire. Wire can be made of lots of strands or of one solid core. Stranded wire is more flexible, but solid-core wire is easier to use. You can also get different thicknesses of wire. For DIY electronics projects such as the ones in this book, I usually use 22 American Wire Gauge (AWG) stranded wire, but it doesn't really matter if your wire is a little bit thicker or a little bit thinner. To use wire to connect your components, you'll need to use a wire stripper. This part will give you plenty of practical experience with this tool, but here are the basics of working with wire and wire strippers.

A wire stripper is an essential DIY electronics tool. Wire strippers do two things: they cut the plastic bit on the outside of the wire, and then they remove the unwanted section of plastic coating. A wire stripper has a few different sizes of holes to fit different sized wires. When you close the wire stripper shut, these holes act like teeth to cut the plastic coating but leave the wire on the inside intact.

To prepare a wire for use with a component, put the wire inside one of the holes on the wire cutter, and close it at the point of the wire where you want to strip the plastic off. With your other hand, hold the wire on the component side. Keep the wire strippers shut, and pull them toward the cut end of the wire. If you've chosen the correct sized hole, this action will cut and remove the unwanted plastic coating. If it doesn't work, try a slightly smaller hole size.

Take the exposed wire, push it through the hole on the component you want to connect it to until most (but not all) of the bare wire is through the hole. Pinch the wire with your fingers so that it bends back on itself, making a loop. Wind the loose end of bare wire tightly around the bare wire on the other side of the loop, making a secure connection to the component to which you want to connect it. You can secure the wire in place by soldering it to your component hole or you can simply wrap a little bit of electrical tape around the loop for a quick, easy fix.

Quick Start Tips

You'll find lots of good tips and tricks for making your first simple robots as you read through this chapter. Here are the most important things to remember:

- *Make sure that your components are wired up securely.*
- *Watch out for loose ends, fraying wires, and untidy connections.*
- *Bare wire is conductive. If two bits of bare wire touch each other, you will cause a short circuit, so strip away only as much of the plastic coating as needed.*
- *Never strip wire that is connected to power. The power sources we use in this book are very small and cannot hurt you, but if you move on to experimenting with robotics that need more power, you should be very careful. Electricity, especially the electricity that comes out of the wall, can really hurt or even kill you.*

Tools and Materials List

To make all the projects in this part, you'll need the following tools:

- Wire strippers
- Wire cutter
- Hot glue gun or superglue
- Scissors
- Craft knife
- Wooden or metal skewer

You'll also need the following materials:

From a Craft Store

- Googly eyes
- Strong craft wire
- Colored felt tip pens
- Ordinary and double-sided sticky tape
- Assorted colors of card
- Paper straw
- Glue stick
- Paper

From an Electronics Store

- Three 3V DC motors
- Three 3V battery packs with on/off switches
- Three batteries
- Stranded electrical wire
- Teknikio sewable light sensor

From Your Home

- Assorted trash, including three bottle corks
- Cardboard box

PROJECT 17

Googly-Eyed Trash Robot

Make your first robot out of trash and a DIY vibration motor

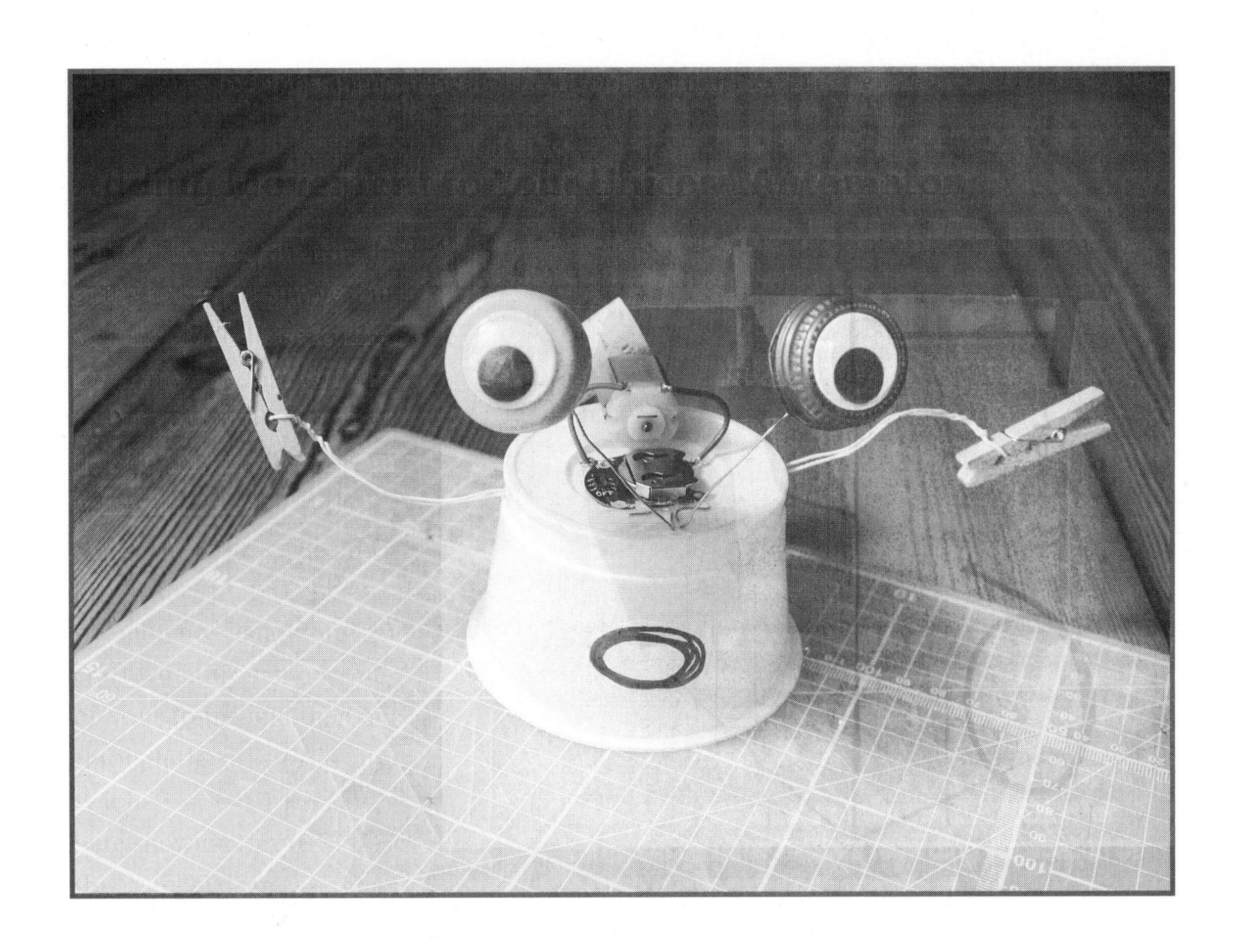

Tools

- Wire strippers
- Wire cutter
- Scissors

Materials

- Googly eyes
- 3V DC motor
- 3V battery pack with on/off switch and battery
- Ordinary and double-sided sticky tape
- Bottle cork
- Craft wire
- Assorted bits of trash

Googly eyes are basically awesome, aren't they? I often carry some self-adhesive googly eyes in my pocket so that I can stick them in places where will they look funny, such as adverts, toys, or furniture. My friend Anders once bought massive googly eyes to stick on his bright yellow tent so that it looked like a giant weird chicken.

Googly eyes can make inanimate objects feel like they have a personality. We're going to use googly eyes to bring a collection of trash to life. We then will make a DIY vibration motor to animate our robot body, completing one of the simplest, cheapest, silliest robots you can make.

This make is for Martin.

Preparing Your Materials

Make sure that you've got all your tools and materials ready. Then go on a trash hunt. You need one cork and two bottle tops, but the trash the rest of your robot is made out of is totally up to you.

Designing your robot out of trash will take a

bit of imagination. Play with different combinations of materials, and think about what personality you want to give your finished robot. Will it be funny, scary, cowardly, weird, or something else?

I wanted to make a noisy robot, so I turned an old yogurt pot upside down for the body. The hollow underside of the yogurt pot amplifies the noises the robot makes when it runs around, making funny clip-clop sounds. I also found some wire and pegs to make into dangly wavy arms, which makes my Googly Eyed Trash Robot look like it is waving its hands around in a panic as it moves.

When you have some trash you'd like to make into a robot, give all your materials a thorough cleaning, and then dry everything. You're now ready to make a DIY vibration motor.

How to Choose and Use Motors

We use motors for movement, from making robot arms wave to driving cars. There are three main types of motors that you might come across: servo motors, DC motors, and stepper motors. We're going to be using DC motors, but it's worth knowing about some of the other motors you might use in the future.

DC motors are good for wheels that need to spin fast and so are perfect for powering wheels to get a robot moving. A *vibration motor* is a type of DC motor with an unbalanced weight on one end. In this project, we'll be making our own vibration motor using an ordinary DC motor and a bottle cork.

Servo motors are fast, powerful, and good for accurate rotation but can only move through 180 degrees. Servo motors are good for robotic arms. *Stepper motors* have slow, accurate rotation with better positional control than servos. Stepper motors can move through 360 degrees and are good for systems where position is very important, such as 3D printers.

Motors can require a lot of power, and some need to be wired up in a special way with other components, so you need to be careful about which ones you buy. In this project, we're using a simple 3V DC motor to make it easy.

How to Use a Wire Stripper

The 3V DC motor that we are using has two wires—positive (red) and negative (black). You should be able to see the colored plastic wire coating and some bare metal wire sticking out at the end. The plastic coating is nonconductive, and the wire inside is conductive. To make a good connection in our circuit, we need to use a wire stripper to expose a little bit more of the conductive metal wire.

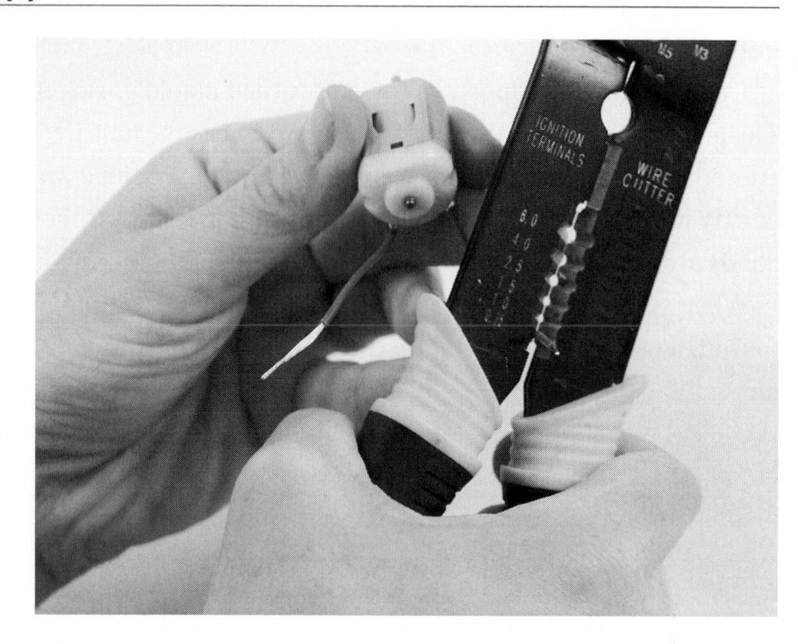

A wire stripper is an essential tool for DIY electronics enthusiasts. Wire strippers do two things: they cut the plastic bit on the outside of the wire, and then they remove the unwanted section of plastic coating. A wire stripper has a few different sizes of holes to fit different sized wires. When you close the wire stripper shut, these holes act like teeth to cut the plastic coating but leave the wire on the inside intact.

Put your wire inside one of the holes on the wire cutter and close it at the point of the wire from which you want to strip the plastic. With your other hand, hold the wire on the component side. Keep the wire strippers shut, and pull them toward the cut end of the wire. If you've chosen the correct sized hole, this action will cut and remove the unwanted plastic coating. If it doesn't work, try a slightly smaller hole size.

Making Your DIY Vibration Motor

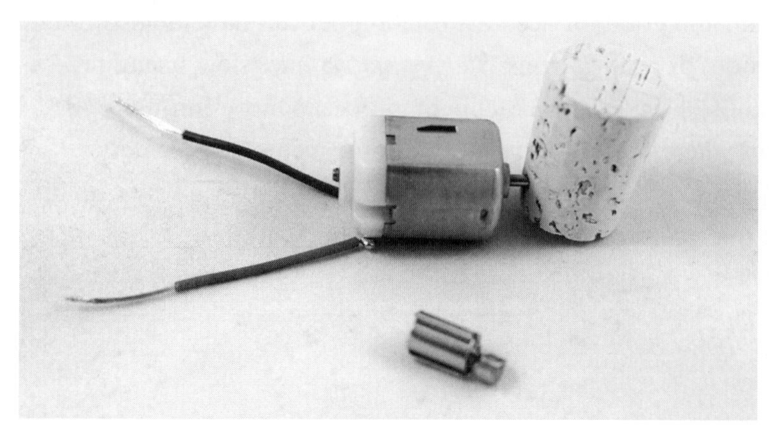

A vibration motor is a type of DC motor with an unbalanced weight on one end. The unbalanced weight makes the motor wiggle all around, vibrating whatever it is attached to. You can buy special vibration motors, such as the tiny one you can see at the bottom of the picture, but for this project we're going to make our own.

Take your cork and push one end of it onto the exposed shaft of the DC motor, just as in the picture. Later, you can experiment by attaching your cork in different positions or by using other bits of trash as your DC motor's unbalanced weight.

Once you've attached your cork securely to the motor, use your wire strippers to expose about ½ inch of wire on each of the leads coming from the 3V DC motor. These bits of exposed wire are what you'll use to make a connection to the 3V battery holder.

Adding Your 3V Battery Holder

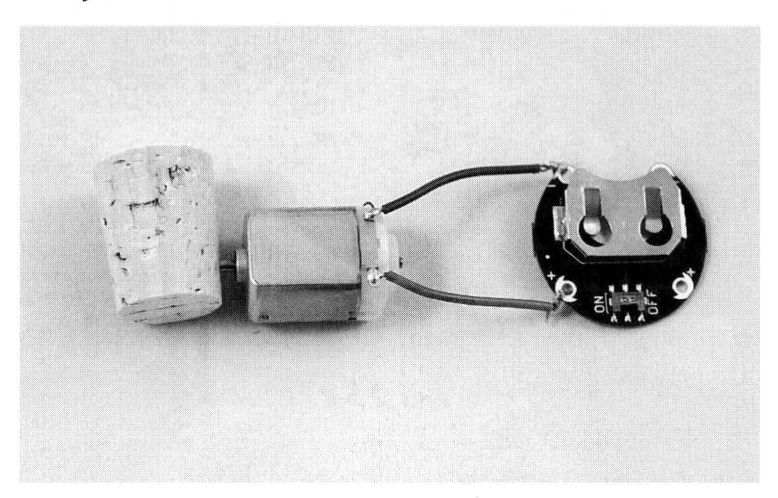

Take the exposed wire on the positive side of your DC motor and push it through the positive hole on your battery holder until most (but not all) of the bare wire is through the hole.

Pinch the wire with your fingers so that it bends back on itself, making a loop. Wind the loose end of bare wire tightly around the bit of bare wire that is close to the battery holder hole so that it makes a secure connection. Repeat this step using the exposed wire on the negative side of your DC motor and the negative hole on your battery holder.

Insert a 3V battery, and briefly switch on the power to check that your circuit works. The end of the DC motor with the cork on it should spin round and round.

Attaching Your Circuit to Your Robot Body

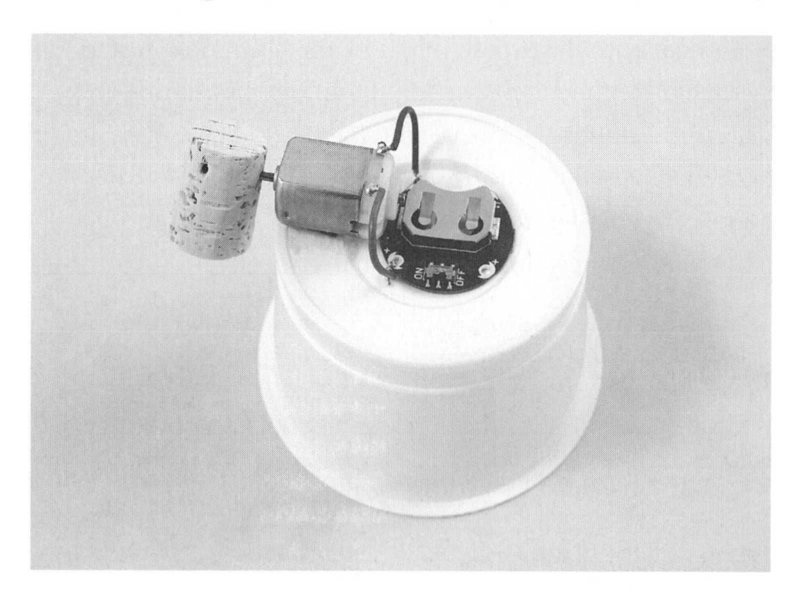

Using double-sided sticky tape or foam, stick the motor in place on your Googly-Eyed Trash Robot. You'll need to put it close to the edge of your robot body, with the cork hanging off the side. Use your finger to turn the cork to make sure that it has enough room to spin without bashing into the body.

Once you are happy with the placement of your motor, use double-sided sticky tape to attach it your 3V battery holder.

You also need to make sure that the weight of the motor doesn't make your robot fall over when the motor is running around. To do this, switch the power on, and give it a try. If your robot falls over, you can try using a different body, changing the placement of the circuit, or adding some weight to the body to give it stability.

Making Googly Eyes on Stalks

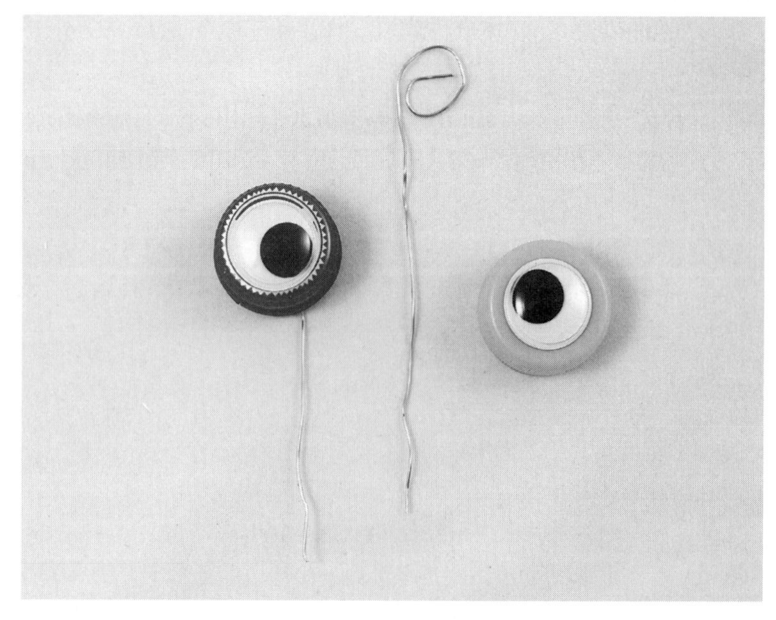

Stick your two googly eyes onto two bottle caps, and then get out your wire cutters and craft wire.

Use your wire cutters to clip off a length of craft wire around 5 inches long. Curl one end of the wire into a flat spiral that will fit neatly inside the bottle cap.

Cut a bit of double-sided sticky tape, and stick it into the middle of the bottle cap. Remove the backing of the tape, and then press the curled-up end of your craft wire stalk onto the double-sided sticky tape on the inside of the bottle cap.

Using your finger, bend the wire so that the spiral is flat at the bottom of the bottle cap and the straight bit of wire bends along the edge of the cap. When the wire reaches the edge of the cap, bend it 90 degrees so that it sticks out like the wire attached to the bottle top on the left of the picture.

Secure the whole thing in place with a bit of ordinary sticky tape, and then repeat the whole process for your other googly eye.

Attaching the Googly Eyes to Your Trash Robot

Once you've made two eyes on stalks, you can attach them to your Googly-Eyed Trash Robot. Place a bit of double-sided sticky tape to the top of your robot body, and then remove the backing.

Take the left eye and bend the craft wire stalk back 90 degrees, about 1 inch from the bottom. Bend the stalk again, this time 90 degrees to the left and about ½ inch from the bottom. This will leave you with a flat corner that you can use to attach the googly eye to your trash robot.

Next, take the right eye and do the same thing, but this time make the second bend 90 degrees to the right. Place both eyes onto double-sided sticky tape on the robot's body, and then use ordinary sticky tape to hold it in place. The wires should look like the picture once they are bent and stuck in place.

Once your eyes feel securely placed, you can bend the craft wire to experiment with different eye positions.

Finishing Your Googly-Eyed Trash Robot

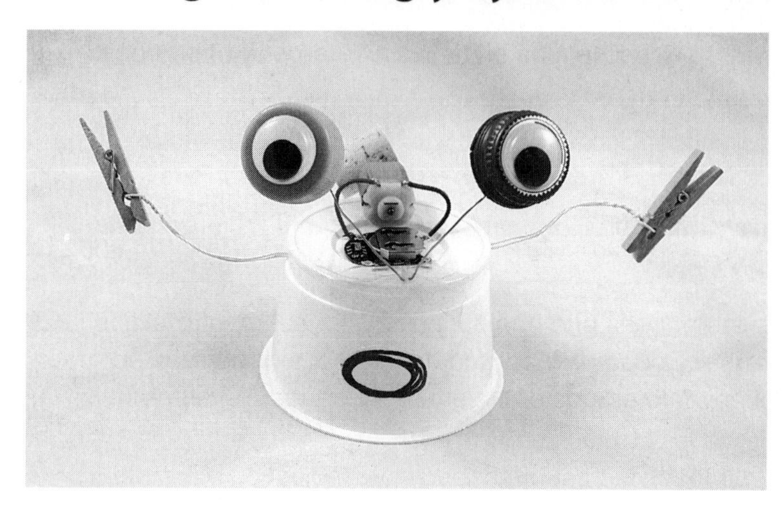

Now that you have your body, motor, power source, and googly eyes in place, you can play around with the look of your robot by adding other bits of trash or craft materials. I added two pegs and some wire to make funny wavy arms. I also drew several different mouths on scrap paper so that I could give my trash robot different emotions to match the eye and hand positions.

Once you've finished your design, flip the switch on your battery holder to make your Googly-Eyed Trash Robot dance. Experiment with your vibration motor by attaching the cork in different positions. Does it change the movement of the robot? How about if you use other things to unbalance the weight of your DC motor instead of the cork?

Fix It

Not working? Don't worry! Follow these steps to figure out why, and fix it.

1. Check your power.
   - Is your battery the right way around? Flip it over and see what happens.
   - Has it run out of juice? Try another battery.
   - Is your battery connecting into your circuit? Make sure that your battery fits snugly into the holder and that the wires are connecting firmly to the correct side of the battery.
2. Check your components.
   - Is your DC motor working? It's a good habit to check each component before you add it into a circuit.
   - Is your DC motor securely fixed in place? Loose connections mean that your circuit won't work.
   - Does your DC motor have an unbalanced weight? If you don't have an unbalanced weight on the shaft of your DC motor, then your robot won't wiggle.
3. Check your wiring.
   - Is your DC motor connected the right way around? The positive wire of your DC motor (the red wire) should be attached to the positive hole on your battery holder,

and the negative wire of your DC motor (the black wire) should be attached to the negative hole on your battery holder.

Make More

If you liked making this trash robot, you can use this guide to make a collection of trash robots with your friends and family and then host a Googly-Eyed Trash Robot sporting competition. You can have a short-distance sprinting race, a wrestling contest, an artistic figure skating dance-off, or a soccer match. Which robot will be the overall winner?

PROJECT 18

Modern Art Robot

Use a DIY vibration motor to make experimental robotic art

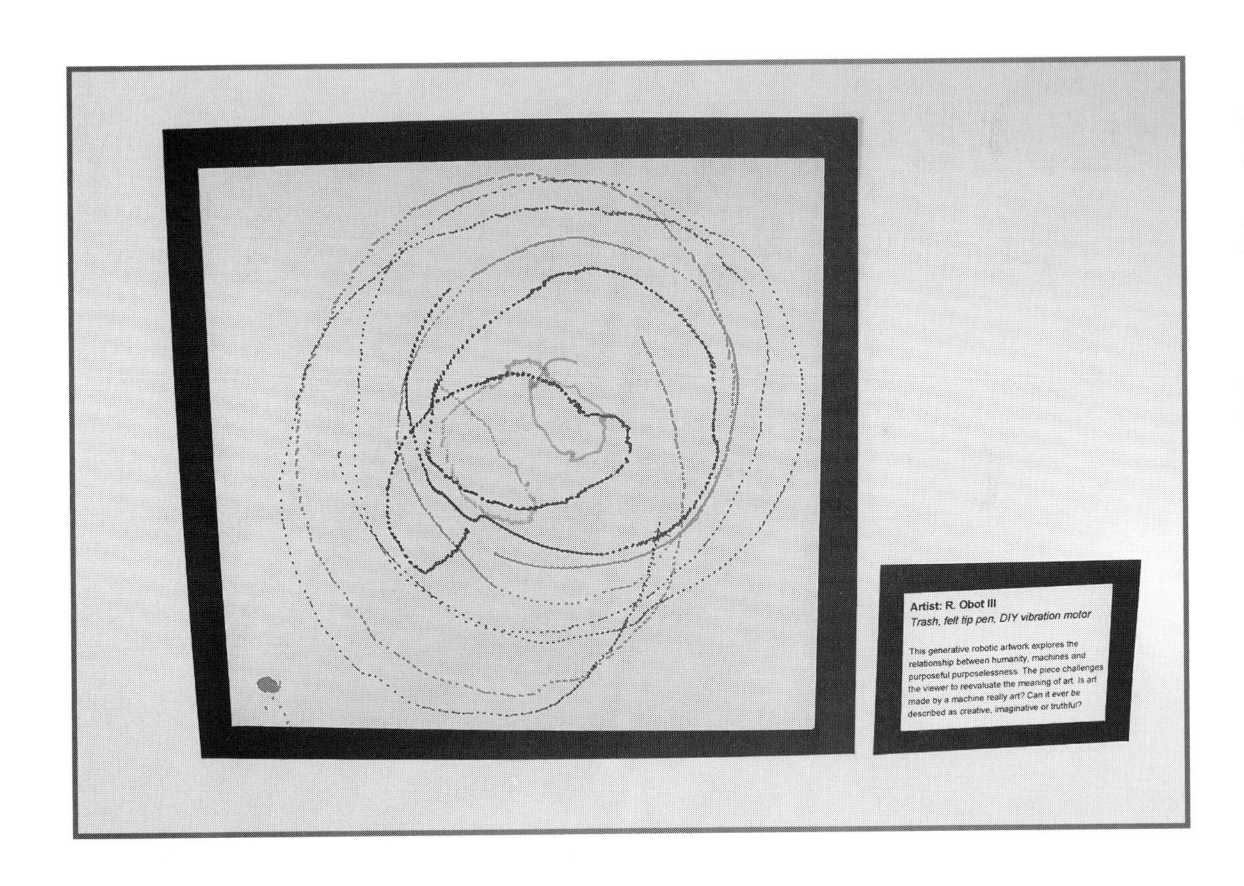

Tools

- Wire strippers
- Wire cutter

Materials

- One 3V DC motor
- One 3V battery pack with on/off switch and battery
- Ordinary and double-sided sticky tape
- Assorted trash, including a cork
- Colored felt tip pens
- Paper

Is art made by a machine really art? Can art made by a robot ever be described as creative, imaginative, funny, or truthful? These might seem like strange things to think about in the context of craft and DIY electronics, but as robots advance, these questions are being asked more and more by engineers as well as philosophers.

In Project 17, we made our own DIY vibration motor. This robot moves in the same way but for a very different purpose. Our Modern Art Robot uses the unpredictable, funny movements of a vibration robot and adds felt tip pen legs that make marks. You can use this robot to make very silly or very serious pieces of art.

This robot is my take on a common robot called a *Scribble Bot*, popularized by a magical place in San Francisco called the *Exploratorium*.

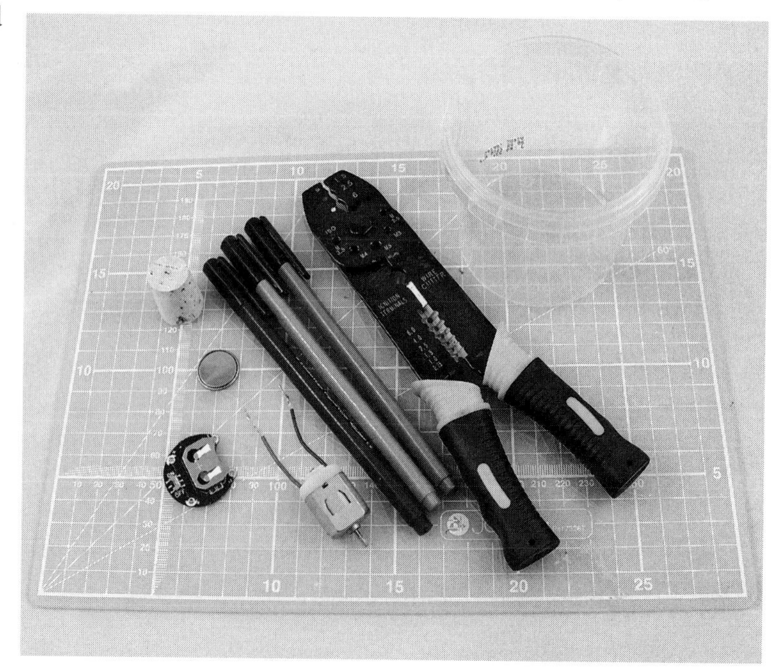

Preparing Your Materials

Make sure that you've got all your tools and materials ready. Then go on a trash hunt. For this make, you'll need a cork and some kind of container to make your Modern Art Robot's body.

I'm using a plastic tub for my robot body, but you

can use an empty tin can, a plastic cup, an egg box, or anything fairly light that you find in the trash. Just make sure that it doesn't have any sharp edges on which you can cut yourself. Once you've found your robot body and a cork, clean them thoroughly and give them a good dry.

Next, you'll need to choose the implements your robot will make its first art with. A good first choice is to use a few colors of felt tip pens, but you can use pencils, chalk, or markers.

The last thing to prepare before you start making is the motor.

Preparing the Motor

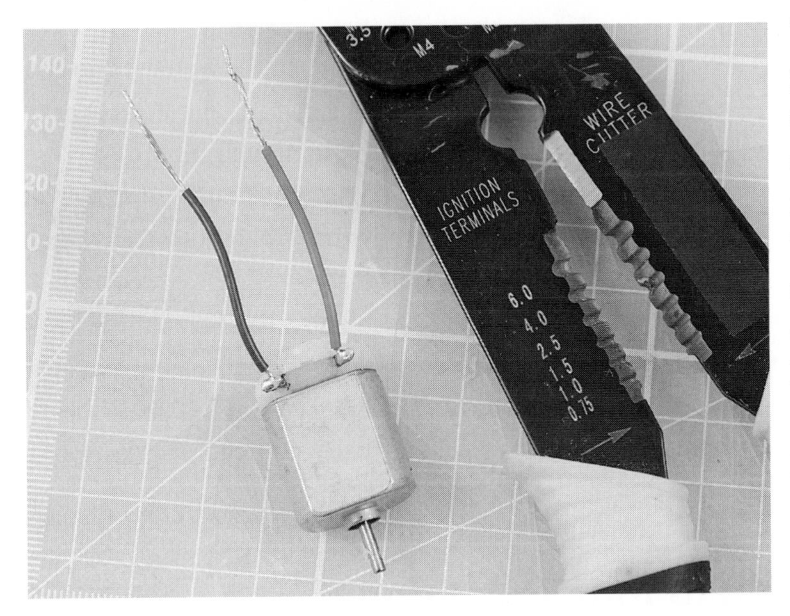

To prepare your motor, use your wire strippers to expose about ½ inch of wire on each of the leads coming from your 3V DC motor. These bits of exposed wire are what you'll use to make a connection to the 3V battery holder.

To use your wire cutter, put one of your wires inside one of the holes on the wire cutter, closing it about ½ inch from the end of the wire. With your other hand, hold the wire on the component side. Don't hold the motor itself because you might damage the connection.

Keep the wire strippers shut, and pull them toward the cut end of the wire. If you've chosen the correct sized hole, this action will cut and remove the unwanted plastic coating. I found that the 1.0-mm hole on my wire cutters worked perfectly for my 3V DC motor wire, but your motor's wire might need a slightly smaller or slightly bigger hole size.

Connecting Your 3V Battery Holder

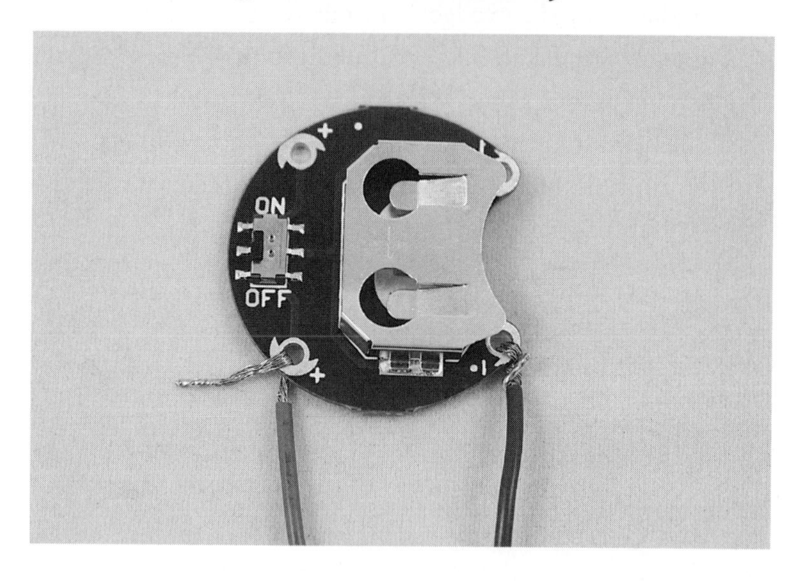

Take the exposed wire on the positive side of your DC motor and push it through the positive hole on your battery holder until most (but not all) of the bare wire is through the hole, just like the wire on the left of the picture.

Pinch the wire with your fingers so that it bends back on itself, making a loop. Wind the loose end of the bare wire tightly around the bit of bare wire that is close to the battery holder hole so that it makes a secure connection. When you're done, it should look like the wire on the left of the picture. Repeat this step using the exposed wire on the negative side of your DC motor and the negative hole on your battery holder.

Insert a 3V battery, and briefly switch on the power to check that your circuit works. The end of the DC motor should spin around.

Attaching Your Circuit to Your Robot Body

Take your cork and push one end of it onto the exposed shaft of the DC motor. Then use double-sided sticky tape to put your circuit in place on the top of your robot body. This cork adds an unbalanced weight to the DC motor, making your Modern Art Robot wiggle around when the battery is turned on.

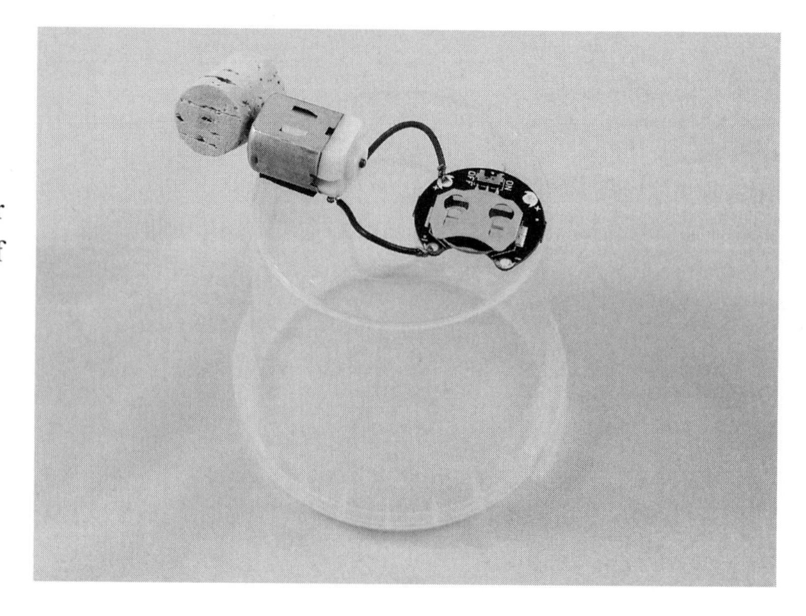

Make sure that you place the DC motor close to the edge of your robot body, with the cork hanging off the side. Your cork will need enough room to spin around.

Your finished robot body should look like the one in the picture. Test it works by turning on the power. The cork should spin around, and the robot body should vibrate and wiggle. Turn off the power—it's time to add the art!

Making the Legs of Your Modern Art Robot

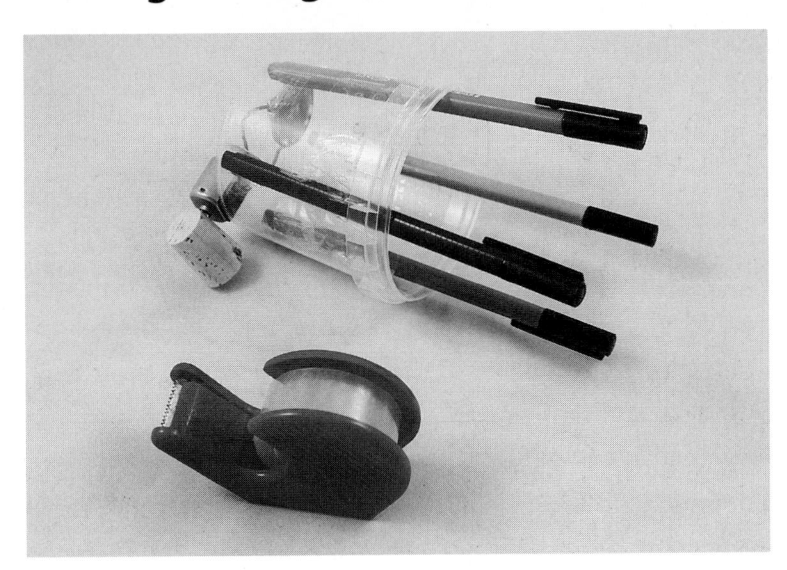

Turn your robot on its side, and get out four felt tip pens and some ordinary sticky tape. Stick one of the pens inside the robot body, aligning the top of the pen with the top of the inside of the robot's body. Use two or three bits of sticky tape to make sure that your robot's leg is secure.

Next, stick a second pen on the opposite side of the body to the first pen in the same way as the first, and then stick the other two pens in between. Once you've given it all four legs, your robot body should look like the picture.

Stand your Modern Art Robot up on its four legs, and check that the weight of the motor doesn't make your robot fall over when it's running around. Switch the power on without taking the lids off the pens and give it a try.

If your robot falls over, make sure that the four legs are taped at the same height, with the tops aligned inside the robot body. You should also check that the spacing of the legs is even.

Preparing Your Canvas

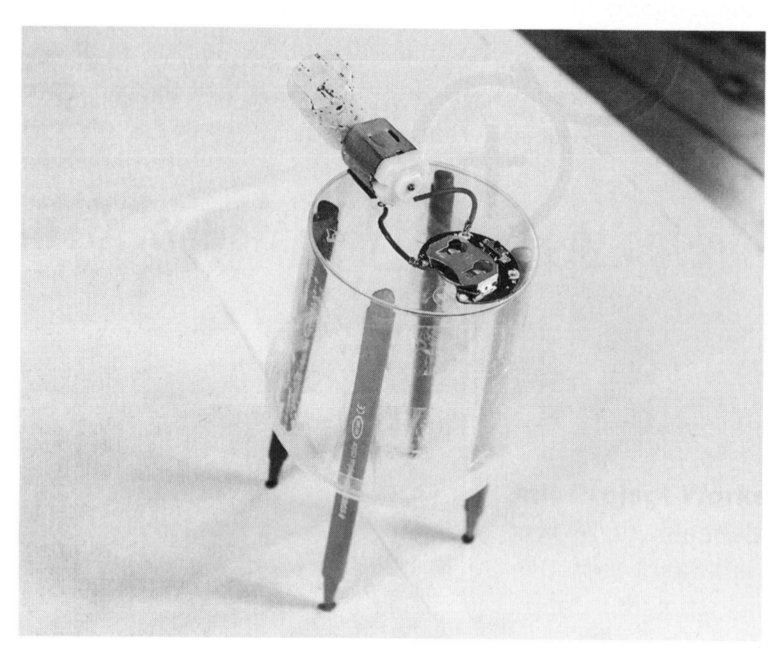

You have finished your Modern Art Robot, but now you need to find a place to create art! To prevent the robot from falling off a table and drawing all over a beautiful carpet, find a wooden or tiled floor, and then lay down something for your robot to draw on.

Your Modern Art Robot will work best on a large piece of paper, so find some poster-sized paper or a big roll of paper or stick a few sheets of normal-sized paper together.

To avoid scribbling all over your floor, limit your robot's range by walling it in with some cardboard or wood. You can even put your robot inside a large cardboard box.

When you have prepared your canvas, remove the lids from your felt tip pen legs, and place your Modern Art Robot onto your paper.

Making Your First Piece of Art

When your Modern Art Robot is finished and your canvas is ready, switch your robot on. The felt tip legs of your creature will support its jittery, wiggly body, making lines, circles, dots, and unpredictable patterns all over the surface of your first canvas.

Take a moment to play with your robot. Give it a

gentle poke or a little turn. What happens? Can you influence the way it travels by moving the paper?

Once you think you can understand the way your Modern Art Robot moves, you can start to work with it to influence the art it makes. Think about the outcome you want, and then experiment with a number of different variables to find out the best way to collaborate with your robot's movements.

How to Experiment with Your Robot's Movement

A *variable* is a thing that you can change. When a scientist does an experiment, they start off by trying to think about all the different things they can change. Then the scientist tries to keep every single variable exactly the same apart from one, which they will change to see what effect it has. This way of working is the basis of what scientists call a *fair experiment*.

You can use this idea to work out the different things you can do to your Modern Art Robot. What effect does having six pens have? How about three? What about if you change the position of your cork, swap the cork for another weight, or remove the cork entirely? You could try making your robot draw on different surfaces, or you could play around with the power by using a run-down battery or even two batteries in series.

Once you have figured out some ways of influencing your robot, you can collaborate with it to design a piece of art. You might choose to use a smaller bit of paper that is absolutely covered in dots and lines and colors, or you might want to use a massive piece of paper with only a few dots. Keep trying and making marks until you have some designs that you like.

Displaying Your Artwork

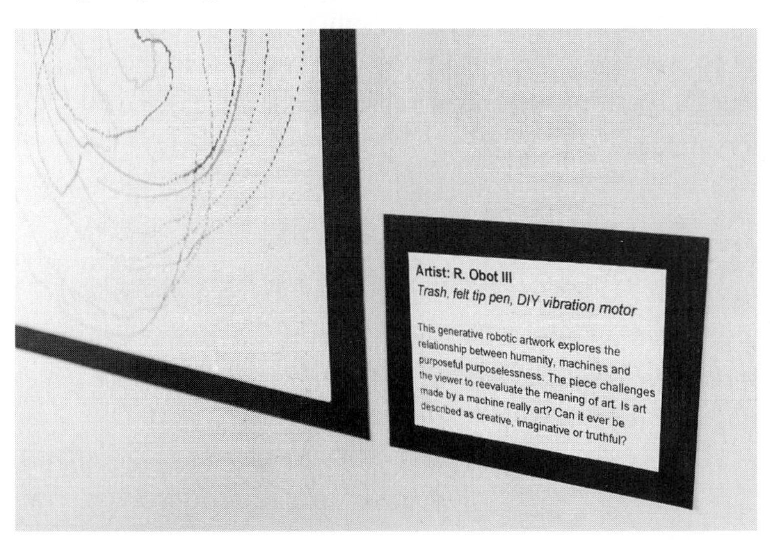

Once you have a few pieces of art that you like, you can hold an extremely exclusive modern art viewing party. To prepare for your party, cut out your favorite designs, and stick them onto some card. This will frame your art and also make it more sturdy so that you can hang it on a wall in an art gallery (or perhaps your bedroom).

Next, you'll need to come up with a description. When you visit an art gallery, there are usually little notices next to each piece of art that tell you who made it and what it is made of. You will also often find a few sentences that tell you what the art is about.

Sometimes these descriptions give historical context, sometimes they tell you about what the artist was feeling, and sometimes they are just pretentious and confusing. Feel free to make your Robot Modern Art description as informative or ridiculous as you like.

Fix It

Not working? Don't worry! Follow these steps to figure out why, and fix it.

1. Check your power.
   - Is your battery the right way around? Flip it over and see what happens.
   - Has it run out of juice? Try another battery.
   - Is your battery connecting into your circuit? Make sure that your battery fits snugly into the holder and that the wires are connecting firmly to the correct side of the battery.
2. Check your components.
   - Is your DC motor working? It's a good habit to check each component before you add it into a circuit.
   - Is your DC motor securely fixed in place? Loose connections mean that your circuit won't work.
   - Does your DC motor have an unbalanced weight? If you don't have an unbalanced weight on the shaft of your DC motor, then your robot won't wiggle.
3. Check your wiring.
   - Is your DC motor connected the right way around? The positive wire of your DC motor (the red wire) should be attached to the positive hole on your battery holder, and the negative wire of your DC motor (the black wire) should be attached to the negative hole on your battery holder.

Make More

My Modern Art Robot used felt tip pens, but you can try using erasable markers on a whiteboard, fat bits of chalk on the pavement, or charcoal art pencils. If you want to make this project into digital art, you can also use glow sticks as legs and download a long-exposure app onto your smartphone to experiment with light painting.

PROJECT 19

Extremely Annoying Robot Alarm Clock

Add a light sensor to your DIY vibration motor circuit to make an autonomous alarm

Tools

- Wire strippers
- Wire cutter

Materials

- One 3V DC motor
- One 3V battery pack with on/off switch and battery
- Ordinary and double-sided sticky tape
- Assorted trash and craft materials, including a cork
- Teknikio sewable light sensor
- Stranded electrical wire (see page 222)

All our robots so far have been controlled by us flipping a switch, but this one responds to a sensor without us getting involved. We're using a light sensor so that the rising of the sun in the morning will activate our Extremely Annoying Robot Alarm Clock.

This robot reacts to a sensor, activating itself without a human turning it on or off. This means that our robot is taking the first steps toward autonomy. An autonomous robot uses sensors to find out about its environment, makes decisions about how to act, and then acts on those decisions without a human. One really cool example of a very complex autonomous robot is a self-driving car. Next time you're in a car, spend some time noticing all the things a self-driving car would have to process, predict, and react to.

Preparing Your Materials

Make sure that you've got all your tools and materials ready. Then search through your trash to find a body for your Extremely Annoying Robot Alarm Clock.

The point of this vibration-based robot is the sound it makes, so spend some time trying out the sound your potential robot bodies will make when they clop and clatter against a surface. Try experimenting with bodies of different sizes. Does a tiny yogurt pot make a bigger sound than a massive soup container? Why do you think this might be so?

As usual, make sure that your chosen trash doesn't have any sharp edges on which you can cut yourself. Once you've found your robot body and a cork, clean them thoroughly and give them a good dry.

Finally, use your wire strippers to expose about ½ inch of wire on each of the leads coming from your 3V DC motor.

Understanding Your Circuit

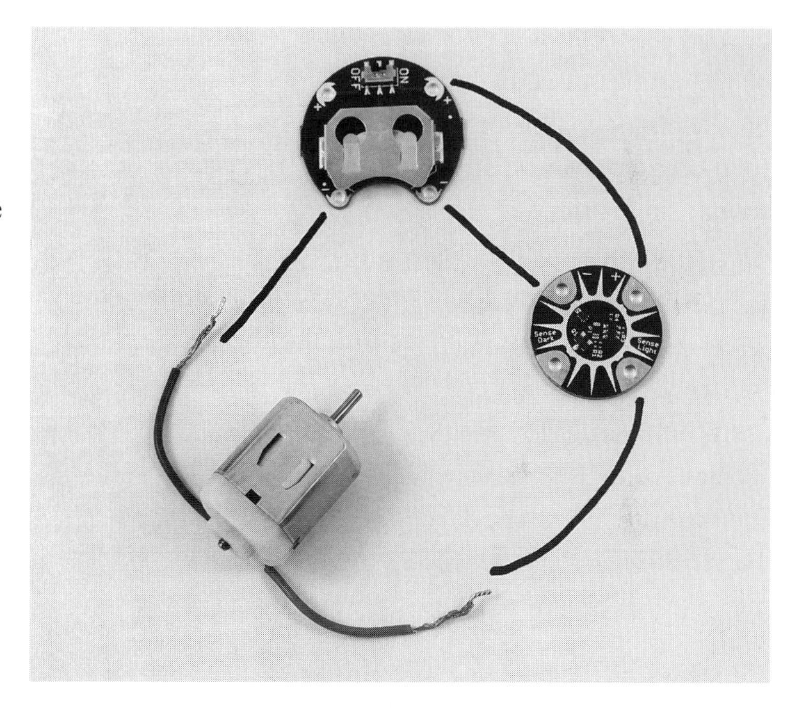

This project uses another DIY vibration motor, but we're going to add another component to the mix—a light sensor. If you made the Dark-Sensing Amulet in the Wearables part (Project 15), you will already be familiar with it.

Take a moment to get to grips with how the components of your circuit will fit together. Your light sensor has four holes, of which we'll be using three. The negative hole (–) needs to connect to the negative bit of your battery holder. The positive hole (+) needs to connect to the positive bit of your battery holder. The hole for "Sense Light" needs to connect to the positive wire of your DC motor (the red wire). Finally, the negative wire of your DC motor (the black wire) needs to connect to the other negative bit of your battery holder.

Make sure that you are familiar with the way your circuit works so that you don't get it wrong later. Look at the picture for a guide to how it all fits together.

How to Choose and Use Different Wires

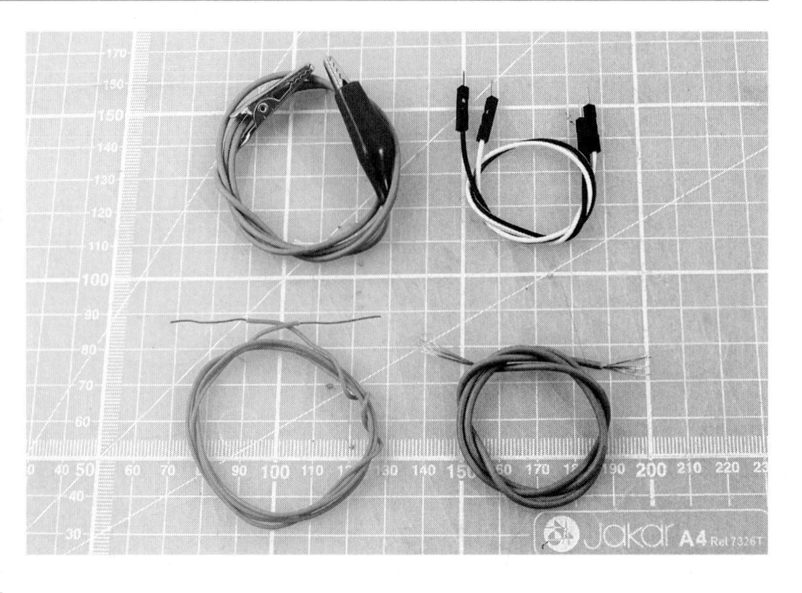

So far in this book we have used copper tape and conductive thread to make most of our circuits. However, sometimes the best option for making a circuit is a wire.

Wire can be stranded (bottom right of the picture) or solid core (bottom left of the picture). If you take off the plastic coating on solid-core wire, you would see one single strand of thick wire. If you take off the plastic coating on stranded wire, you would see lots of smaller bits of wire. Stranded wire is much more flexible, but solid core wire is easier to use. You can also get different thicknesses of wire. This thickness is measured in either millimeters (mm) or wire gauge (AWG).

For DIY electronics projects such as the ones in this book, I use stranded wire 22 AWG thick, but it doesn't really matter if your wire is a little bit thicker or a little bit thinner.

The other wires in the picture are crocodile clips (top left) and jumper cables (top right). We're not using these in this book, but they will be very useful for prototyping your own DIY electronics projects in the future.

Wiring Up Your DC Motor

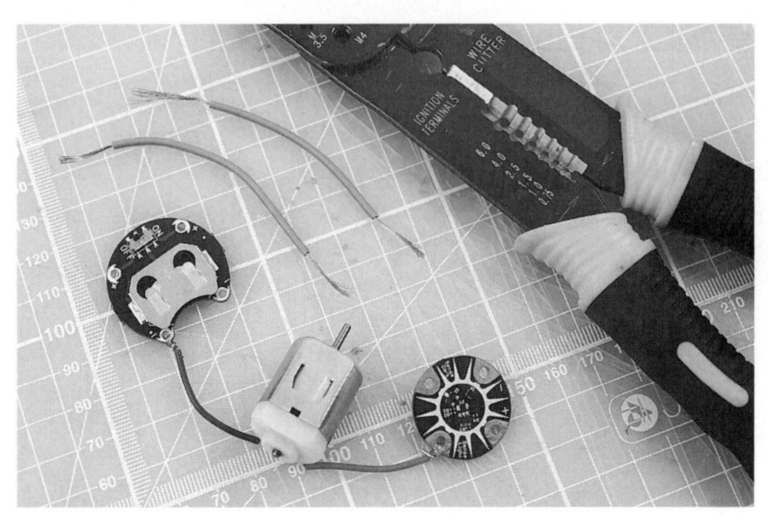

Take the exposed wire on the positive side (the red wire) of your DC motor and push it through the "Sense Light" hole on your light sensor. Then make a loop and wrap the bare wire tightly around itself to make a secure connection between the wire and the hole.

Now take the exposed wire on the negative side (the black wire) of your motor and push it through one of the negative holes on your battery holder. Make a loop with the wire, and wrap it tightly around itself.

Next, prepare the stranded wire that you're going to use to connect the positive and negative sides of the light sensor to your 3V battery pack. Use your wire cutters to cut two lengths of stranded wire, one around 3 inches long and the other around 4 inches long. Finally, use your wire strippers to expose about ½ inch of wire on both ends of the two lengths of stranded wire.

Completing Your Circuit

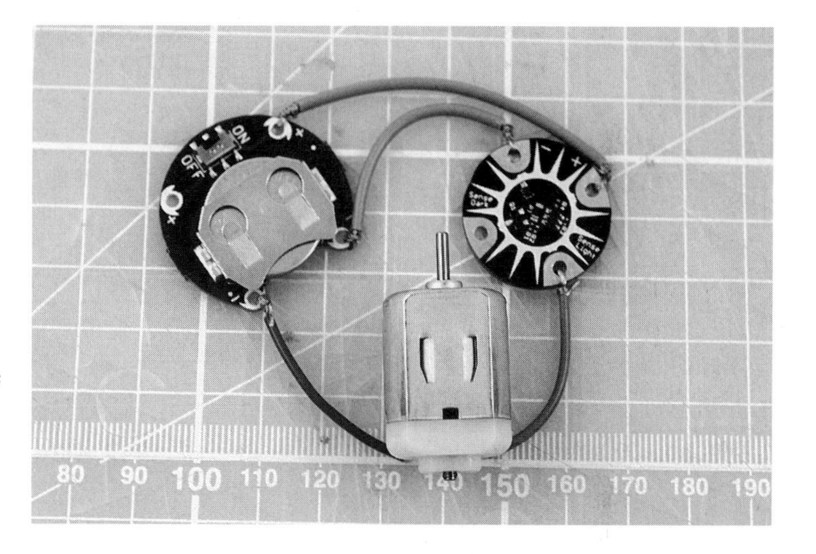

Take the 3-inch-long piece of stranded wire and push one end of the exposed wire through the negative hole on the light sensor. Make a loop with the wire, and wrap it tightly around itself. Take the other end of this wire and push it through the negative hole on the battery pack. Again, make a loop with the bare bit of wire, and wrap it tightly around itself.

Now take the 4-inch-long piece of stranded wire and push one end of the exposed wire through the positive hole on the light sensor. Then wrap it round itself. Take the other end of this wire and push it through the positive hole on the battery pack before making and tightening the final loop.

Your twisted connections should hold firmly enough to make a circuit, but if you want to make them extra secure once you've checked that they work, you can keep them in place with some hot glue or electrical tape. If you know an adult with a soldering iron, you should try to convince them to show you how to use this awesome DIY electronics tool to melt metal and really secure your connections at each of the six points in this circuit.

Attaching the Circuit to Your Robot Body

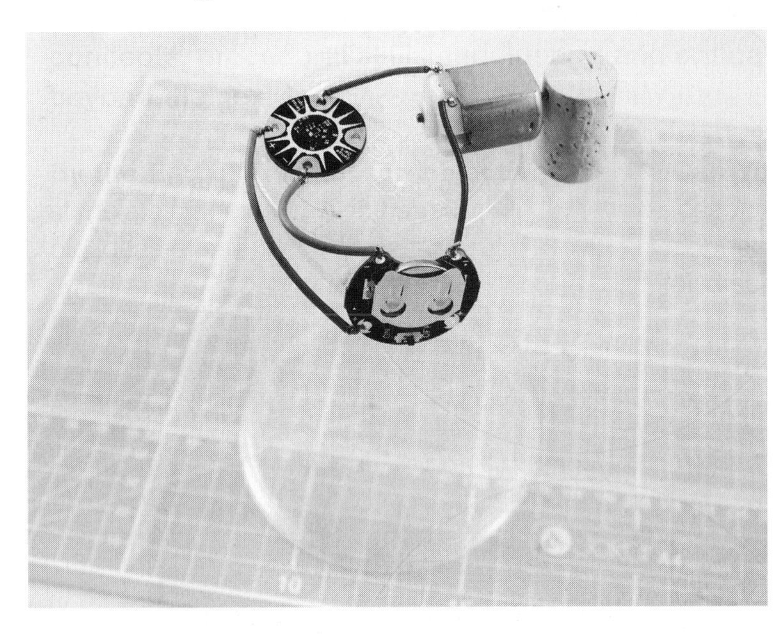

Push one end of your cork onto the exposed shaft of the DC motor, and then use double-sided sticky tape to put your circuit in place on the top of your robot body. As usual, place the DC motor close to the edge of your robot body, with the cork hanging off the side, so that your cork has enough room to spin around.

Your light sensor should rest on the top of your Extremely Annoying Robot Alarm Clock. Depending on the size of your robot body, you may need to arrange your circuit so that the battery pack sticks to the side rather than the top.

Your finished robot body should look like the one in the picture. Test that it works and that the weight of the motor doesn't make your robot fall over by turning on the power. The cork should spin around, making the robot body vibrate and wiggle without falling on its side.

Making Your Robot Alarm Clock Extra Annoying

Your basic robot body is now finished, but you need to think about how to make it super annoying so that you make sure that you get out of bed to turn it off.

To do this, you can use anything clanky, jingly, or loud. Bottle caps, bells, or anything metal works well for this project. You can also add a false floor inside your container and fill it with beads, rice, or dried beans to make a minishaker.

If you want an extra DIY electronics challenge, you could also try to hook up a sewable buzzer that triggers at the same time as your vibration motor. Try adding the buzzer to the light sensor in series with the motor. Does it work? How about if you add it in parallel? Do you think that the battery is powerful enough to power the motor and the buzzer?

Finishing Your Robot Alarm Clock

Attach your annoying noise-making implements to the body of your Robot Alarm Clock. I chose to use some jingly metal bells suspended on long bits of string so that when my robot wakes up, the bells smash into each other and make an extremely annoying sound!

To use your Extremely Annoying Robot Alarm Clock, you'll need to flip the battery switch on at night when it's dark. Your light sensor won't be able to tell the difference between natural light and artificial light, so you'll have to turn it on when you're about to turn off your overhead lights too.

Place your activated Extremely Annoying Robot Alarm Clock somewhere near a window, and then, when the sun comes up in the morning, it will turn on and start making a racket to wake you up. Remember that your robot will move around while it is turned on, so make sure that it has plenty of space in which to jump around.

Fix It

Not working? Don't worry! Follow these steps to figure out why, and fix it.

1. Check your power.
   - Is your battery the right way around? Flip it over and see what happens.
   - Has it run out of juice? Try another battery.

- Is your battery connecting into your circuit? Make sure that your battery fits snugly into the holder and that the wires are connecting firmly to the correct side of the battery.

2. Check your components.
   - Are all the parts of your circuit working? It's a good habit to check each component before you add it into a circuit.
   - Are all the parts of your circuit securely fixed in place? Loose connections mean that your circuit won't work. If you're having trouble making the wires connect firmly enough, ask an adult to help you twist them on, or try soldering the connections in place.
   - Does your DC motor have an unbalanced weight? If you don't have an unbalanced weight on the shaft of your DC motor, your robot won't wiggle.
3. Check your wiring.
   - Is your DC motor connected the right way around? The negative hole of your light sensor needs to connect to the negative bit of your battery holder. The positive hole of your light sensor needs to connect to the positive bit of your battery holder. The hole for "Sense Light" needs to connect to the positive wire of your DC motor (the red wire). Finally, the negative wire of your DC motor (the black wire) needs to connect to the other negative bit of your battery holder.

Make More

Making robots respond to sensors is a lot of fun. When designing your own autonomous robotic creations, my advice is to keep it simple to start with. Design your DIY robots with one input (the thing that triggers something, such as a sensor or a switch) and one output (the thing that does something such as lights, sound, or movement).

For more complex robots, you'll need to add a "brain" and write instructions for your robot using code. There are lots of great beginner microcontroller "brains" made especially for robotics. Check out Crafty Robot's SmartiBot, Adafruit's Crickit, or the BBC Micro:bit to take your robotics journey to the next level.

PROJECT 20

Unicorn Automaton

Engineer cardboard with a low-tech make of my favorite mythical beast

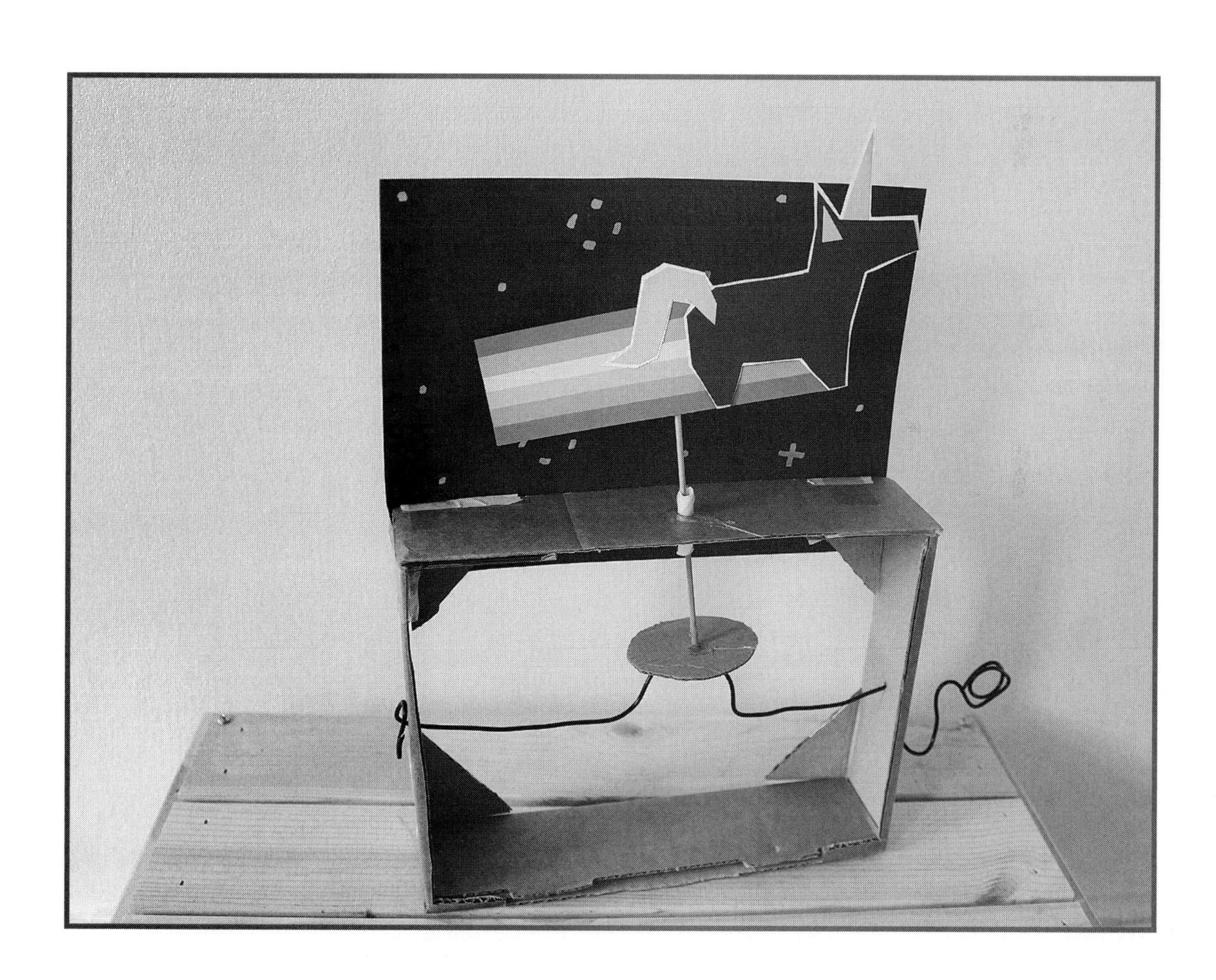

Tools

- Wire cutter
- Hot glue gun or superglue
- Scissors
- Craft knife
- Wooden or metal skewer

Materials

- Cardboard box
- Ordinary and double-sided sticky tape
- Strong craft wire
- Assorted colors of card
- Paper straw
- Glue stick

The last make of this book is a paper craft engineering experiment. This project does not have any electronics in it. I've left them out on purpose so that you can think back over skills you've learned and the projects you've made over the course of this book and then add your own DIY electronics.

An *automaton* is a machine or robot designed to do a sequence of things. This project is a very simple automaton, but you can find beautiful examples of extremely complex automatons dating back as far the 1700s.

My automaton is inspired by unicorns. I love making electronics with a playful or silly twist and one of my most popular makes of all time was a Gesture Controlled Robot Unicorn. I made my first Robot Unicorn out of cardboard, tin foil, two cheap motors and a tiny computer called a Micro:bit. Once you've made your own Unicorn Automaton, you can think of how you will make your creation unique using craft, electronics, or even code!

Preparing Your Materials

Make sure that you've got all your tools and materials ready. Then search for the perfect box. My box was 8 inches long, 6 inches wide, and 3 inches deep, but yours doesn't have to be exactly the same dimensions.

Turn to pages 256–257 for templates of all the parts you'll need to make your Unicorn Automaton. You can use what ever colors you like for the body and head of your unicorn.

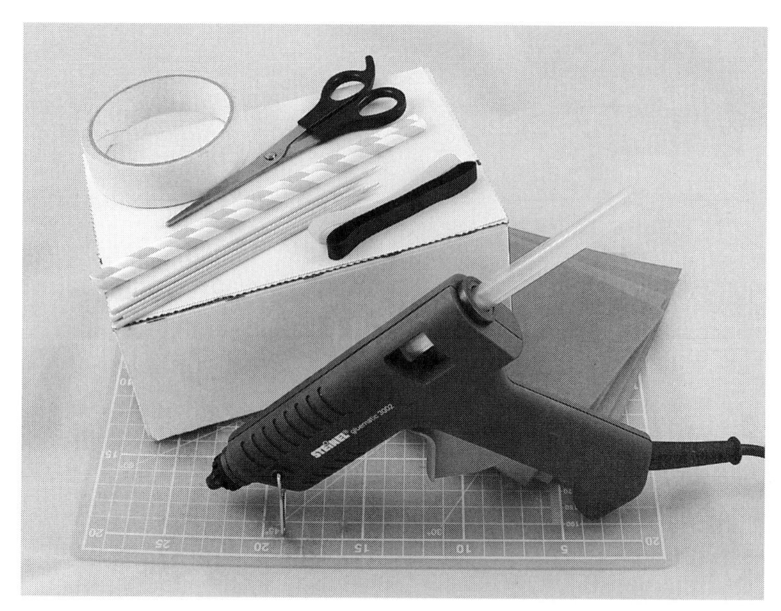

Your automaton doesn't need to be a unicorn—you can make any real or mythical creature, from your favorite Internet cat to Pegasus, the flying horse from Greek mythology.

To make the rainbow you'll need red, orange, yellow, green, blue, and purple colored card.

You'll also need to cut out the large circle on the template page and four or five smaller circles. You can make these circles using scrap cardboard left over from making your frame in the next step.

Making Your Frame

Turn your cardboard box on its side, and use a pair of scissors or a craft knife to carefully cut off all the flaps, setting the spare bits of cardboard aside for later. This rectangle will be the frame for your Unicorn Automaton.

Once you have your cardboard frame, give it a poke on one edge. You'll notice that it's not super stable, so we need to add some corners to keep it from collapsing.

Use some of the scrap cardboard to cut out two squares measuring 2 inches on each side. Cut both the squares diagonally in half, leaving you with four right-angled triangles.

Use sticky tape to secure a triangle in each corner of your box, just as in the picture. Give your cardboard frame another poke. Adding those four corners should have made your frame way more stable.

Making Holes in Your Frame

Next, you need to make three holes for the Unicorn Automaton parts to fit into your frame. Use a ruler to find the horizontal and vertical center of the panel on the left-hand side, and then mark it up with a pencil. Do the same thing for the panel on the right-hand side, and then measure and mark the panel on the top of the frame.

Take your wooden or metal skewer and use it to pierce through the holes you marked on both sides of the frame, just as in the picture. Remove the skewer once you've made the holes.

Now use your wooden or metal skewer to push through the hole you marked at the top of the frame. Then use a pencil to make the hole a bit bigger. Cut a 1-inch piece off your paper straw, and then insert it halfway into the larger hole at the top of your frame.

Set your frame aside, and move on to making your unicorn.

Making Your Unicorn Automaton

Lay out your unicorn template pieces. Use a dab of glue to stick the unicorn horn onto the head of the unicorn body template. Next, glue the body and horn of the unicorn to a piece of white card. Glue the smaller inner ear triangle to the larger outer ear triangle then stick the ear onto the head of your unicorn behind the horn.

Next, add on your tail. I've included a simple tail template but you can use your craft skills and imagination to make your own version. How about a tail made of tinsel, feathers or crepe paper? However you choose to make your tail, stick it on with glue.

Finally, cut around the white card surrounding your finished unicorn, leaving a small margin of card that will help make the colors stand out. This backing card will also make your unicorn sturdier.

Finishing Your Unicorn Automaton

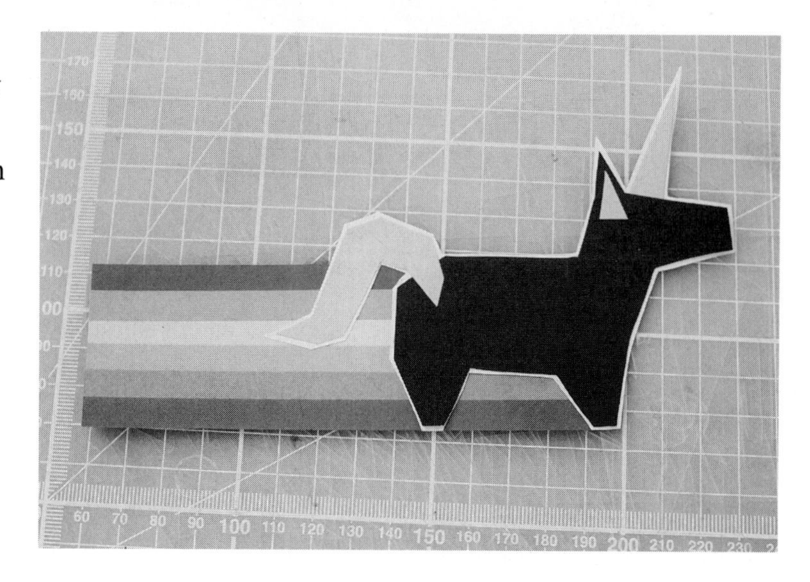

The final thing to make for your unicorn is the rainbow trail behind the flying pop-tart kitty. You can do this in two ways.

The easier way is to cut a 1.5- × 5-inch strip of white card, and then use a ruler and a pencil to divide this strip into six long, thin segments. You can then use pencils, pens, or paint to color the rainbow.

The way I did it is slightly more time-consuming, but it results in a very brightly colored tail. I made six strips of card in red, orange, yellow, green, blue, and purple and then glued them in a stack on top of each other, starting with red and ending with a small strip of purple. I then used my craft knife and a ruler to trim the whole thing to 1.5 by 5 inches.

Once you've finished your rainbow trail, glue one end of it to your unicorn, just as in the picture. I also added a final strip of card on the back of the unicorn just to make sure that the rainbow was securely attached to the unicorn's body.

Assembling Your Automaton

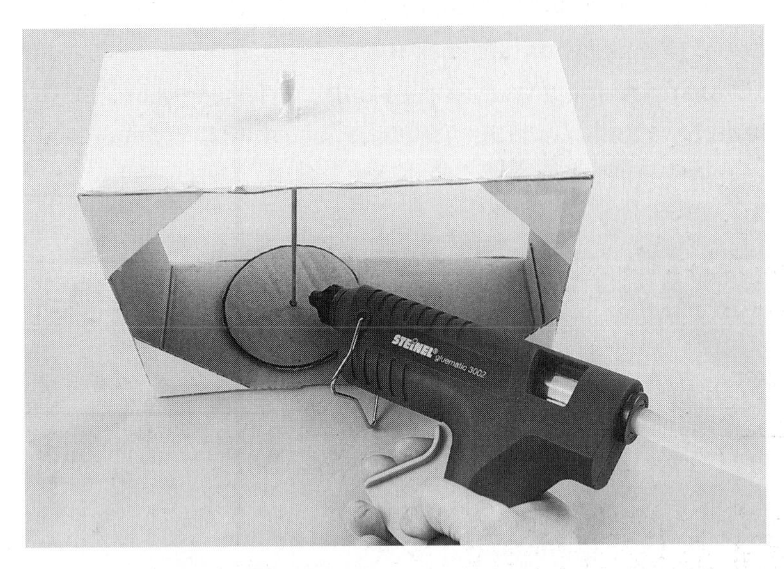

Now that you have your frame and your unicorn body, you're ready to start assembling your automaton. This automaton will move a central stick up and down. This central stick is where we will be attaching the unicorn body so that it looks like it is flying through space when the stick moves up and down.

If you have one, heat up your glue gun. If you don't, you can use superglue. Whichever glue you're using, you should find an adult to help you with this bit of the make so that you don't burn yourself with the hot glue or stick your fingers to your face with the superglue.

First, glue the bit of paper straw to the frame so that it doesn't move around or fall out. Next, stick a wooden skewer through the straw. This skewer will be your central stick, and the straw will help to keep it in place. Take the large circle you cut out of cardboard earlier and stick the skewer right through the centre of the circle. Use your hot glue or superglue to stick the circle securely in place. Then trim off any excess skewer poking out underneath the circle.

The thing you've just made is called a *cam follower*. Read on to find out more about cams and cam followers and the cool things you can do with them.

How to Use Cams in Automatons

A *cam* is a part of a machine that either rotates or moves backward and forward. The movements of the cam make the cam follower react in certain ways.

You can make your DIY automatons do loads of different movements using different shapes and placements of your cams and cam followers. The mechanism I've used in the next step is an extremely simple up-and-down motion, but if you want to, you can experiment with different motions using the smaller cardboard circles you cut out earlier.

Push a straightened length of craft wire through the sides of your frame in the way I describe in the next step. However, before you push your wire through the other side of

the frame, push one of the smaller cardboard circles onto the wire. This is your cam. By placing it in different positions in relation to the cam follower, altering the shape, pushing the wire through in different locations, or adding more than one cam, you can experiment with automaton movement.

If you want to learn more about cams and movement, look up the Exploratorium's excellent guide to DIY automatons.

Adding Movement to Your Unicorn Automaton

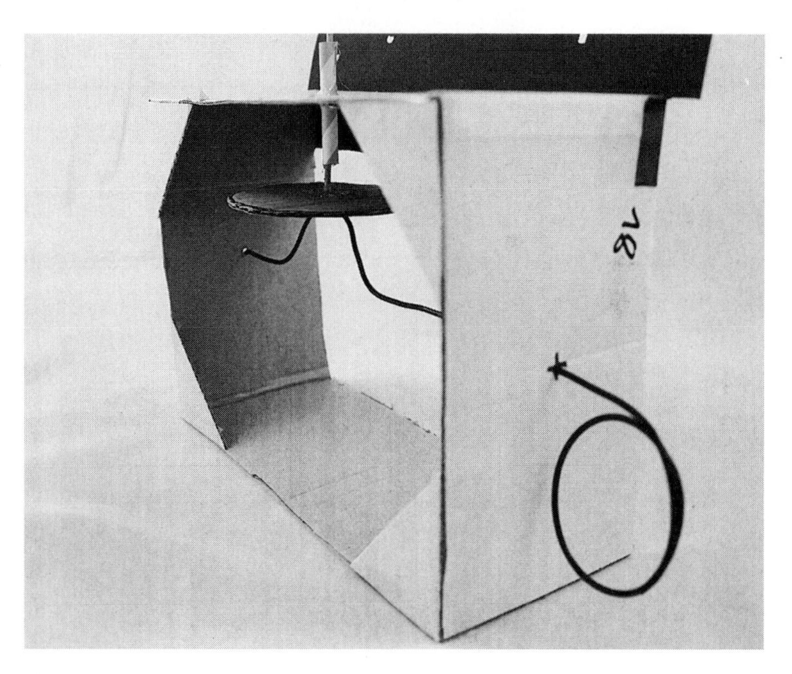

For our first automaton, I'm going to keep it simple, but you should feel free to mix it up in any way that you like.

Put your frame down on its side, and push your cam follower and central stick out of the way, toward the top of the frame. Push a straightened length of thick craft wire through one side of your frame and then out through the other side, until you have about 1 inch of wire showing through. Bend this wire 90 degrees so that it lies flat against the side of the frame.

Next, pull through a little more wire from the other side, and use your fingers to make a curve underneath the cam follower, just as in the picture. Cut the other end of the wire about 4 inches away from the frame, and then use your finger to bend it into a circular handle.

Put your frame the right way up again. The cam follower should fall down onto the curved bit of wire. Turn the handle, and your wire should push the cam follower and central stick up and down. When the curve is pointed down, the cam follower should rest on the straight wire. When the curve is pointed up, the cam follower should rest on the top of the curve. You may need to adjust your curve to get the right movement.

Finishing Your Unicorn Automaton

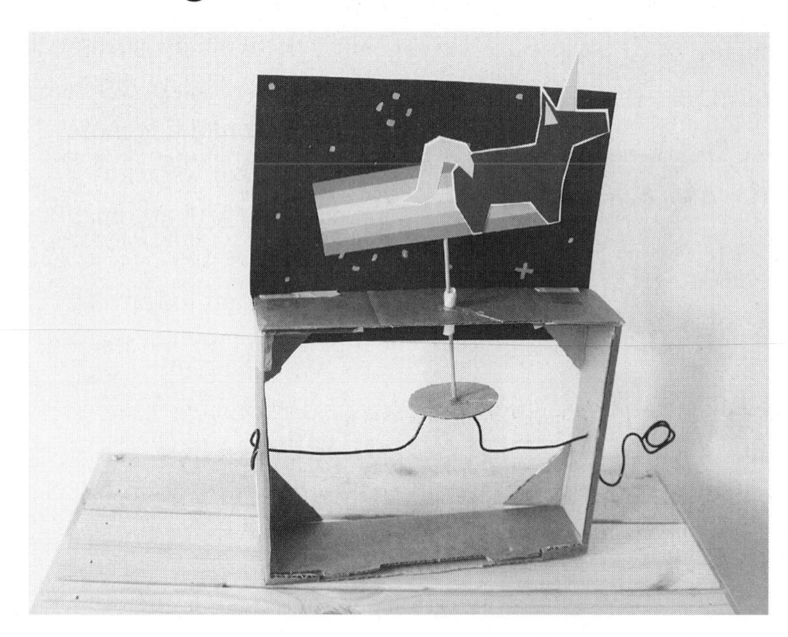

Now that all your unicorn Automaton mechanisms are in place, you can finish your design. Use sticky tape to secure your Unicorn onto the central stick of your automaton. Give your handle a few turns to see if the unicorn moves in the way that you were expecting. Adjust the angle of your wire if you need to.

Finally, add in the background. Choose a piece of colored card and cut it to size. It should be about 6 inches high and about an inch wider than your box. Once you're happy with your background, stick it onto your frame by cutting two tabs on either side, then bending them round and sticking them to the sides of the frame with tape.

That's it! Happy flying! Take a look at the Make More box below to see what's next.

Make More

You may have noticed that the Unicorn Automaton project doesn't have any electronics in it. This is on purpose. Instead of me telling you how use to electricity to make this project even cooler, I want you to think about all the different ways you could use the DIY electronics skills that you've learned over the course of this book. Use the "Maker Notes" pages at the end of this book to write down some ideas and work out what you need to do next.

Good luck, and have fun!

Maker Spotlight

Name: Phoenix Perry
Location: London
Home: New York
Job: Founder of Code Liberation Foundation, game developer, and lunatic extraordinaire

Phoenix Perry creates physical games and user experiences with hardware and sound. As an advocate for women in game development, she founded Code Liberation Foundation, an organization that teaches women to program games for free. In her role at Goldsmiths, University of London, she lectures on physical computing and games and leads an MA program in independent games and experience design. Her research looks at using our senses to play games, with a particular focus on sound- and skin-based feedback to trigger responses.

Q. ***What do you like to make and craft?***
A. I make games and playful installations with hardware and sound. Generally, I use a microcontroller paired with a game engine such as Unity 3D or a creative coding platform such as Processing or Open Frameworks. My materials are often wood, plastic, conductive materials, hacked game controllers, motors, sensors, cameras, and vinyl. By expanding the field of video games to include the physical world and our bodies, I have a richer canvas on which to solder and design. I'm exploring new ways our senses can be used in play, and I'm particularly interested in creating communities of players using touch and sound.

Q. ***What are you currently working on?***
A. Right now, I am working on two games, Bot Party and Thrum. Also, I'm planning some workshops for the nonprofit I founded called Code Liberation. Code Liberation catalyzes the creation of digital games and creative technologies by women, nonbinary, femme, and girl-identifying people to diversify science, technology, engineering, art, and mathematics (STEAM) fields. At the moment, I am focusing on mentoring women who want to teach people about machine learning using play.

Q. ***What are you doing next?***
A. I want to take a vacation and not work on anything at all. I feel like I want some time to dream and stare at clouds. After that, I want to see if I can take some of the

lessons I learned about vibration from Thrum and add them to Bot Party to make the game better.

Q. ***What is your dream project?***
A. Right now, I'm dreaming of making a vibrating floor for people to just lie on in a gallery. I am inspired to play with generative patterns, specialized sound, and vibration.

Q. ***Who inspires you?***
A. Octavia Butler, Donna Haraway, Ursula LeGuin, Sherry Huss, Sophie Kravitz, Adelle Lin, Helen Leigh, Sara Ahmed, Lauren McCarthy, and so many other amazing women who create futures. Fiction inspires me. My favorite books at the moment are *Who Fears Death* by Nnedi Okorafor, *Parable of the Sower* by Octavia Butler, *The Great Derangement* by Amitav Ghosh, *Children of Time* by Adrian Tchaikovsky, and *The Dispossessed* by Ursula Le Guin.

Q. ***Can you share your favorite websites or places you go to learn new skills?***
A. Some of my favorite websites for tutorials and inspiration are Kobakant, Adafruit, and Hackaday. My absolute favorite book is *The Art of Electronics*.

Q. ***What is your favorite place to get materials or tools?***
A. I love the website www.lessEMF.com for anything to do with e-textiles and soft circuits.

Bot Party

Bot Party, a game I am working on with Charlie Ann Page, is an interactive sound experience for humans. These three bots have a problem. They have no way to communicate with each other, but you can help repair their broken network. They need you to touch another human holding a bot. Using your skin like cables, the bots use

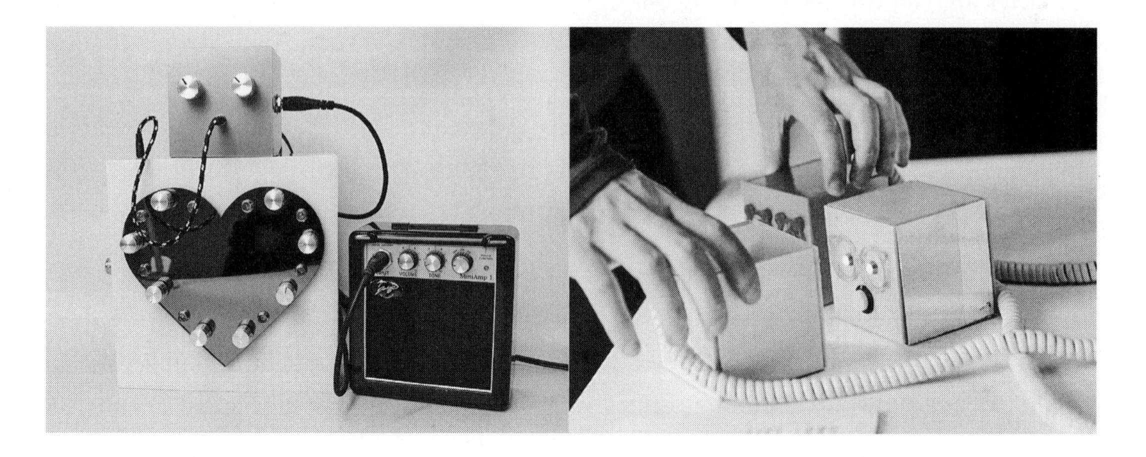

the proprietary bot-to-skin-to-skin-to-bot communication protocol (BSSB) to send encoded secret messages to each other. Hold hands with other players to get the bot sound spectacular started!

Bot Party has left the prototype stage, and now I am designing a connected mobile app. Right now it is debuting at games festivals around the world. I have so many ideas for levels I want to try out.

Thrum

Thrum, a game I prototyped with Heather Kelley, is inspired by bees. Players crawl around on a 250- × 250-cm hexagonal wooden floor trying to find a lone vibrating bee and then follow her instructions to visit a chosen flower in the environment. Thrum is only a prototype right now, and it might end up becoming a launching pad for another project. It's okay to make things that change a lot or don't work out at all. It's part of the learning process. If you want to make games, it's important to workshop ideas and see whether you can get anything meaningful out of them or whether they need to go back to the drawing board.

Twitter: @phoenixperry and @playbotparty
Websites: Phoenixperry.com, Playbotparty.com, Touchinteraffective.com

PART FIVE

Templates

Maker Notes

Project:____

My ideas:____

Materials needed: ____________________________________

Tools needed: ______________________________________

Notes for next time: __________________________________

Maker Notes

Project:____

My ideas:____

Materials needed: ____________________________________

Tools needed: ______________________________________

Notes for next time: _________________________________

Maker Notes

Project: ______________________________

My ideas: ______________________________

____

____

____

____

Materials needed: ______________________________

____

____

____

____

Tools needed: ______________________________

____

____

____

____

Notes for next time: ______________________________

____

____

____

Maker Notes

Project:____

My ideas:____

Materials needed: ____________________________________

Tools needed: ______________________________________

Notes for next time: __________________________________

Maker Notes

Project:____

My ideas:____

____

____

____

____

Materials needed: ____

____

____

____

____

Tools needed: ____

____

____

____

____

Notes for next time: ____

____

____

____

Maker Notes

Project:____

My ideas:____

Materials needed: ____________________________________

Tools needed: ______________________________________

Notes for next time: __________________________________

Templates

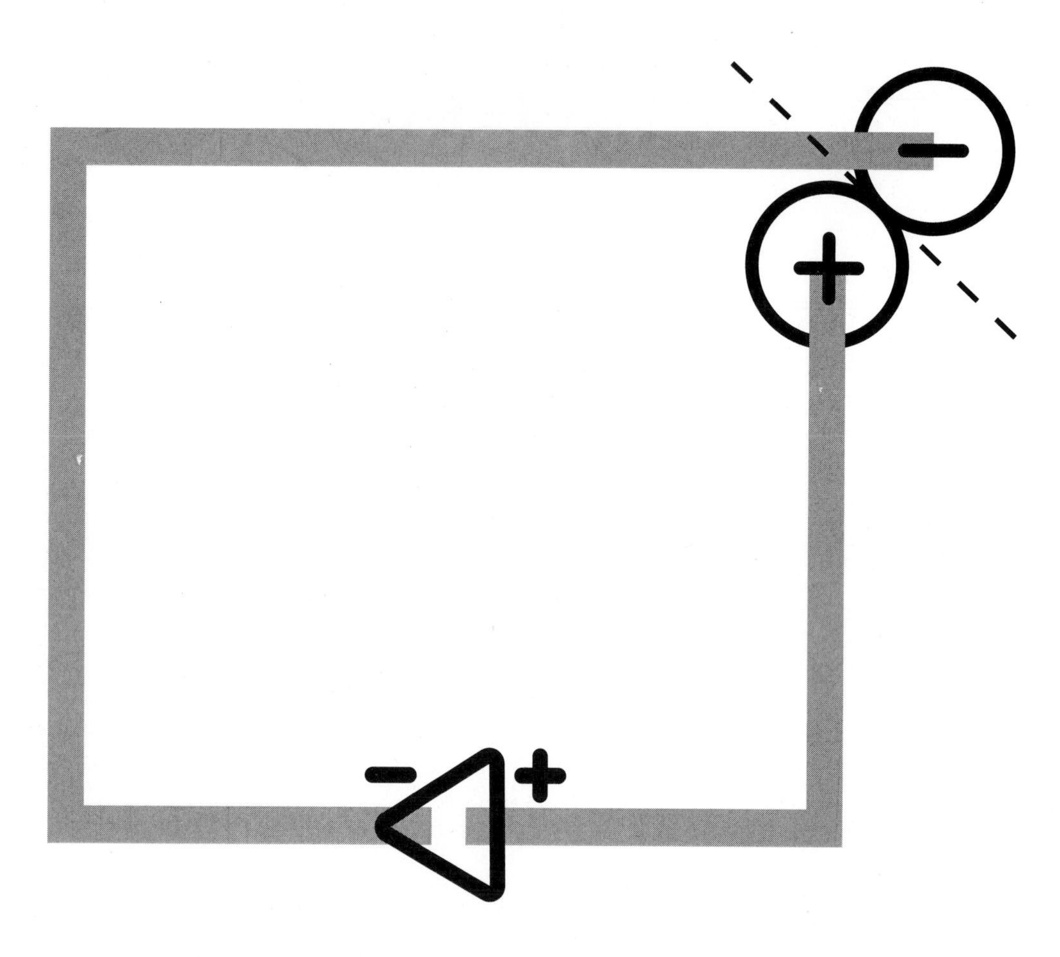

PROJECT 1: Light-Up Greeting Card

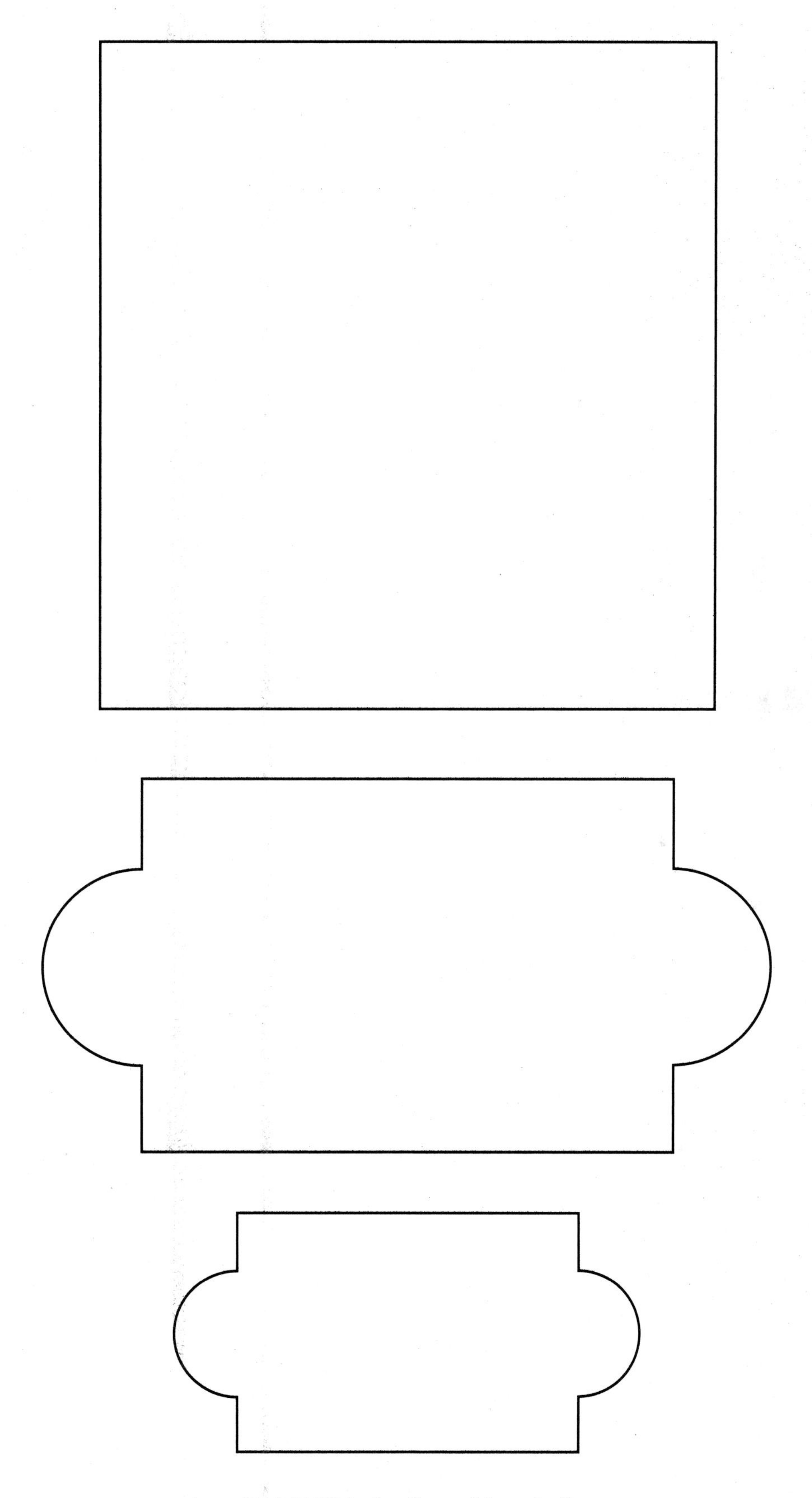

PROJECT 3: Cardboard Doorbell

PROJECT 8: Squishable Sparkle Heart

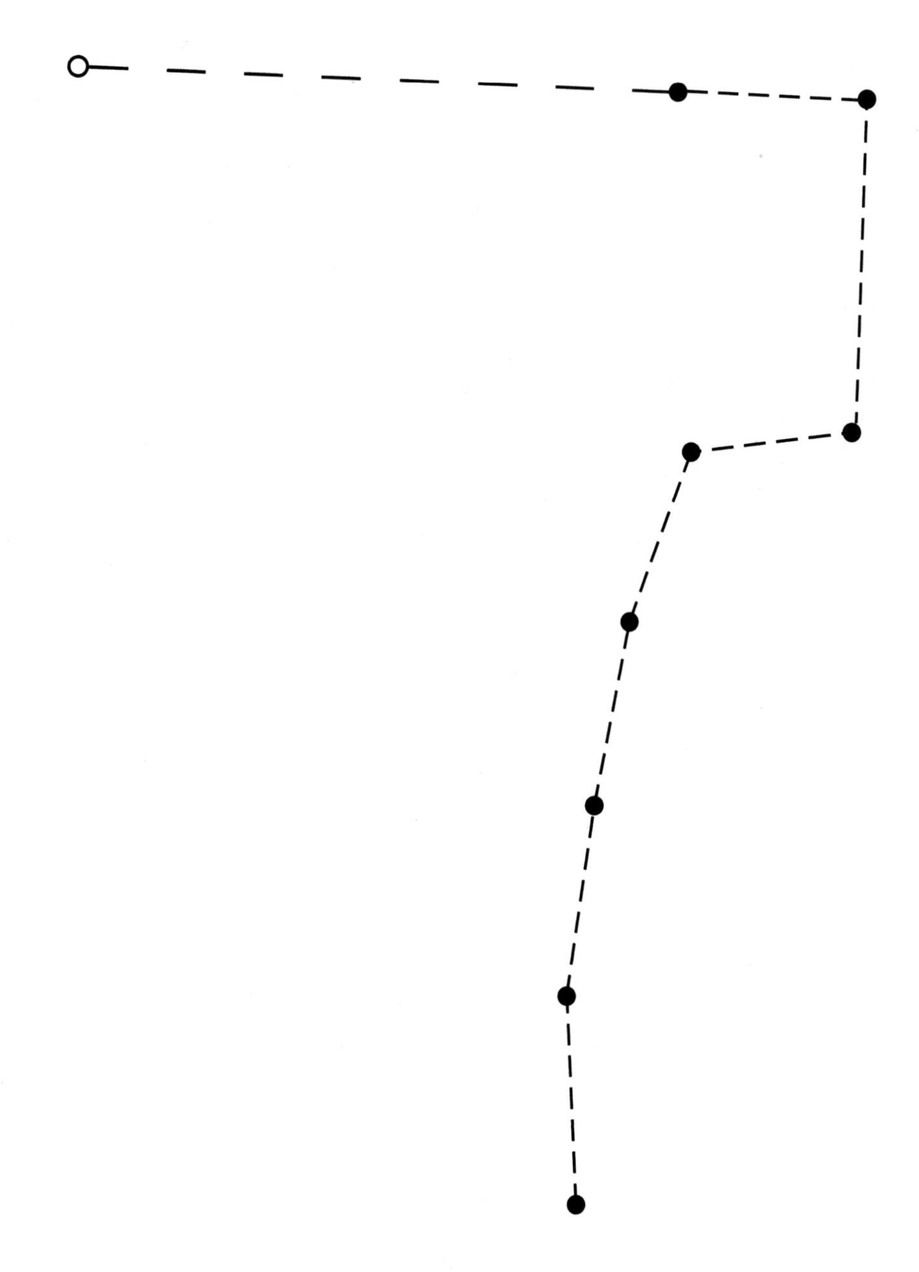

PROJECT 10: Constellation Night Light

PROJECT 9: Tiny Squishy Torch

PROJECT 11: Grumpy Monster with DIY Tilt Sensor

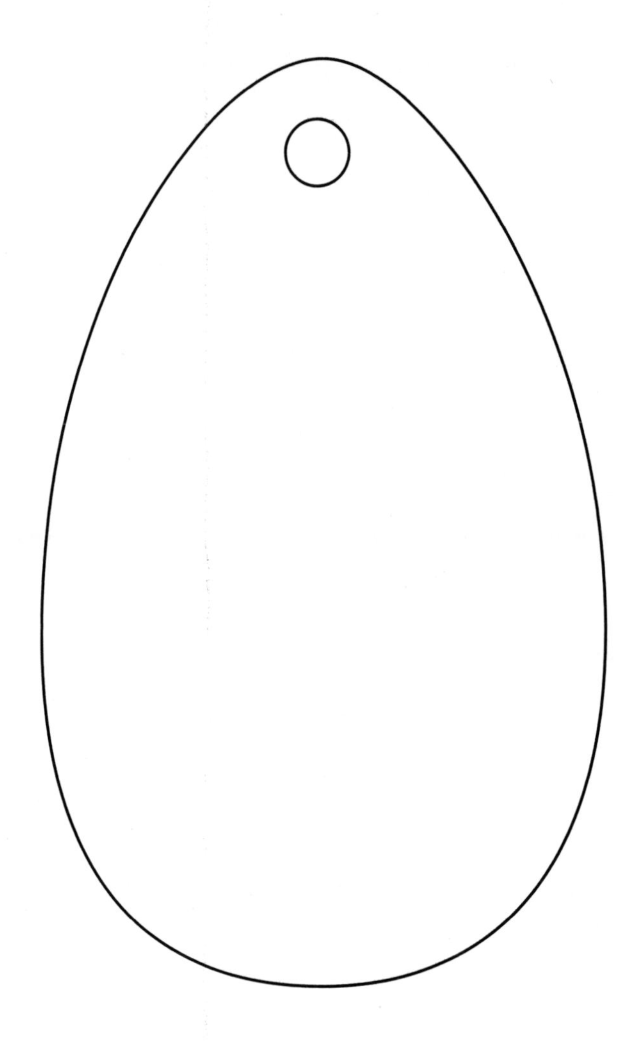

PROJECT 15: Dark-Sensing Amulet

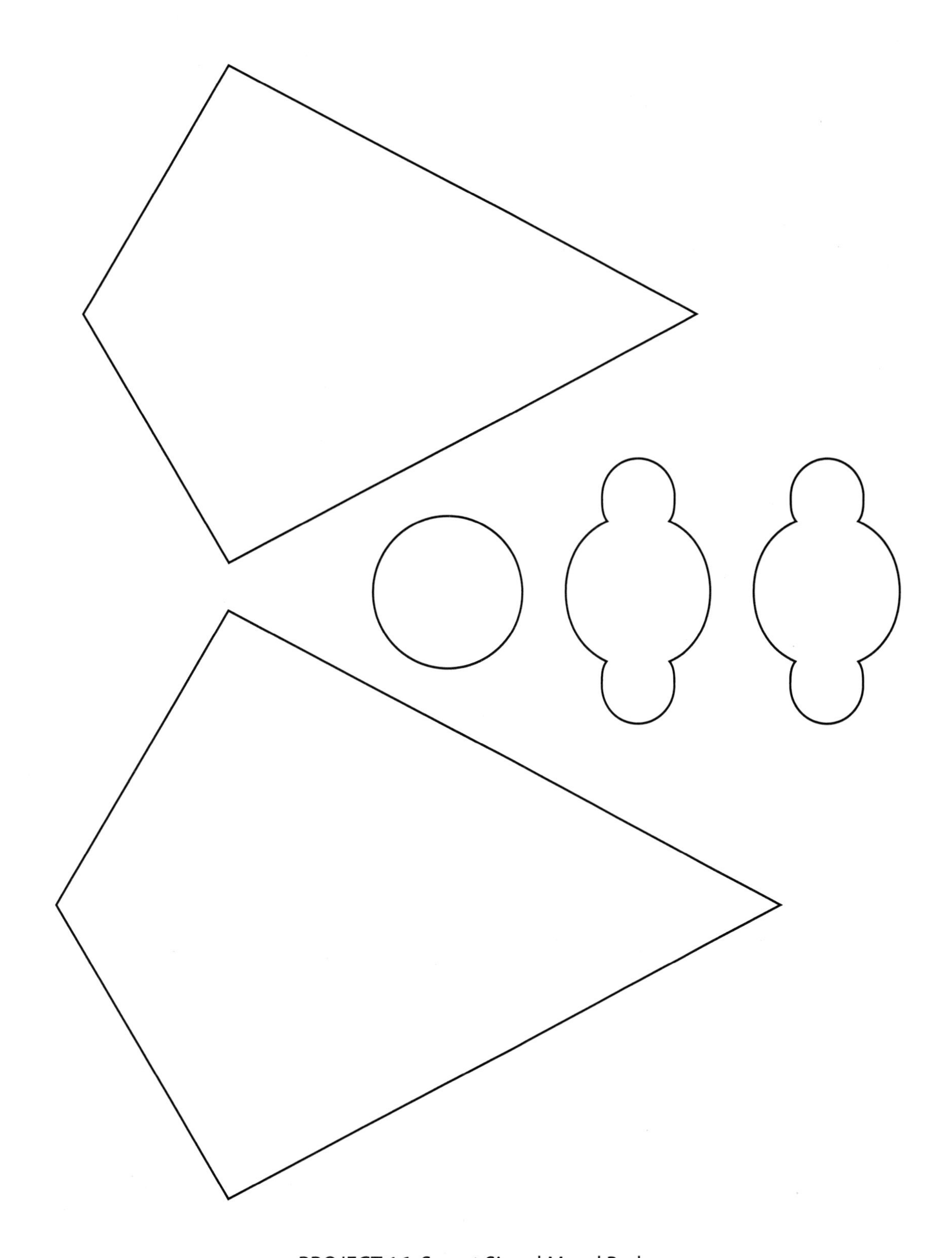

PROJECT 16: Secret Signal Mood Badge

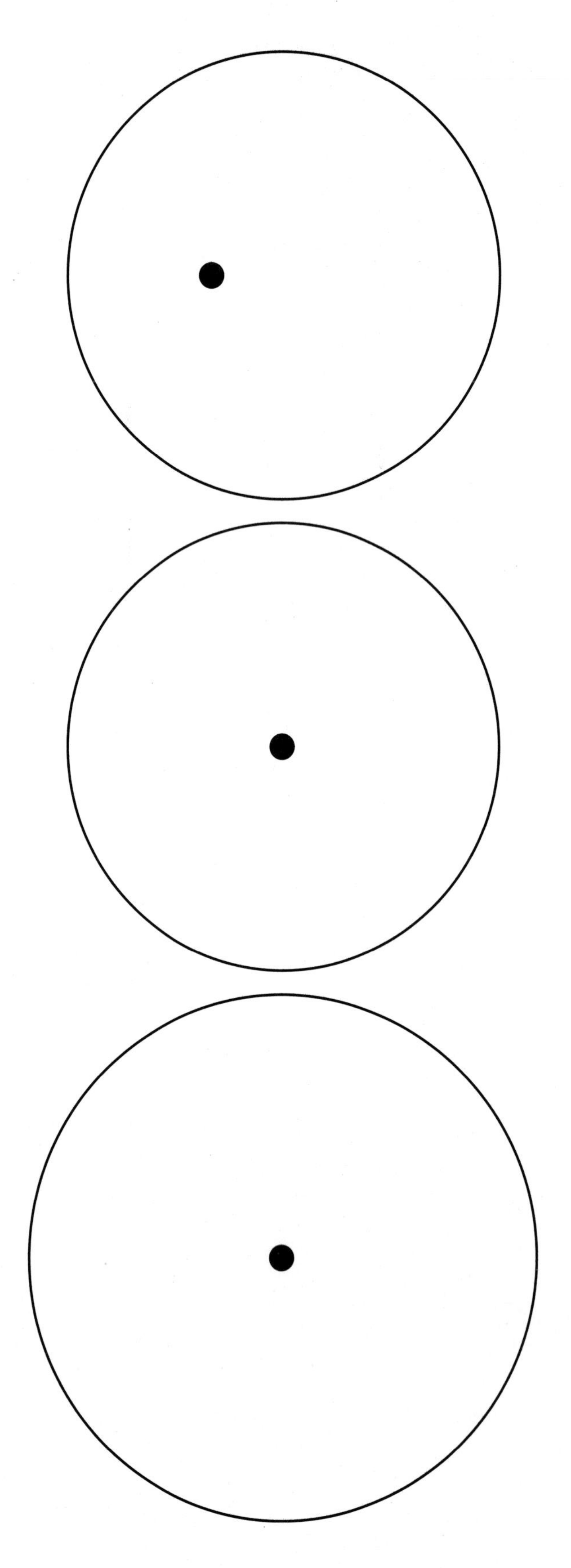

PROJECT 20: Unicorn Automaton

PROJECT 20: Unicorn Automaton

Index

L

M

N

O

P

Q

R

U

V

W